The Finite Element Method in
Partial Differential Equations

The Finite Element Method in Partial Differential Equations

A. R. Mitchell
Department of Mathematics, University of Dundee

and

R. Wait
*Department of Computational and Statistical Science,
University of Liverpool*

A Wiley—Interscience Publication

JOHN WILEY & SONS
Chichester · New York · Brisbane · Toronto

Library of Congress Cataloging in Publication Data:

Mitchell, Andrew
    The finite element method in partial differential
equations.

    'A Wiley—Interscience publication.'
    Includes bibliographical references.
    1. Differential equations, Partial—Numerical solu-
tions. 2. Finite element method. I. Wait, R.,
joint author. II. Title.
QA377.M572   515'.353   76-13533

ISBN 0 471 99405 7

Typeset in IBM Century by Preface Ltd, Salisbury, Wilts
and printed in Great Britain by The Pitman Press, Bath

To
Ann and Fiona

Preface

There is no longer any need to sell the finite element method as a technique for solving partial differential equations. This is particularly so in the case of elliptic equations where at the moment it has taken over from the finite difference method. It is a good example of a topic which transcends many boundaries and its development has only been made possible by cooperation between engineers, mathematicians and numerical analysts. Because of the breadth of interests of its devotees it is easy to convince oneself that there is not a suitable text on the finite element method, a point of view which has led to a rapidly growing literature on the subject. The material in the present book is intended to bridge the gap between the well known works of Zienkiewicz (1971) and Strang and Fix (1973), which represent the finite element interests of engineers and mathematicians respectively. At no time do the present authors take sides in the long-standing controversy regarding the relative merits of finite difference and finite element methods. It is sufficiently gratifying to know that two such powerful techniques exist for the numerical solution of partial differential equations.

Most of the book is aimed at final-year undergraduate and first-year postgraduate students in mathematics and engineering. No specialized mathematical knowledge is required for understanding the material presented beyond what is normally taught in undergraduate courses on vector spaces and advanced calculus. An exception to this is Chapter 5, which can be omitted on a first reading of the book. Hilbert space and functional analytic concepts are introduced throughout the book mainly from the point of view of unifying material. Only a working knowledge of partial differential equations is assumed since anything beyond this would seriously limit the usefulness of the book. Since a variational principle rather than a partial differential equation is often our starting point, a chapter on variational principles is included with suitable references to more advanced works on the subject.

We hope that practical users of the finite element method will also find the book useful. For their benefit we have covered as many variants of the finite element method as possible, viz. Ritz, Galerkin, least squares and collocation, and in Chapter 4 we have given a large selection of possible basis functions to be used with any of the above methods. To balance this overcoverage of material in particular areas we have omitted eigenvalue problems. Our reason (or excuse) is that these are more than adequately covered in Chapter 6 of Strang and Fix (1973).

The list of references is restricted to those texts actually referred to in the book. For a more complete list of references see *A Bibliography for Finite Elements* by Whiteman (1975). Some recent texts and conference proceedings devoted mainly to finite element methods are listed in the references for the convenience of interested readers. These are Zienkiewicz (1971), Aziz (1972), Oden (1972), Strang and Fix (1973), Gram (1973), Lancaster (1973), Miller (1973, 1975), Whiteman (1973, 1976), Watson (1974, 1976), De Boor (1974), Oden, Zienkiewicz, Gallagher and Taylor (1974), and Prenter (1975).

Much of the material in this book has been presented in the form of lectures to Honours and M.Sc. mathematics students in the Universities of Dundee and Liverpool. Also at the invitation of the Institutt for Atomenergi the former author lectured on the material of Chapters 2 and 4 at the Nato Advanced Study Institute held in Kjeller, Norway in 1973, and the latter author lectured on the material of Chapters 3, 5 and 6 at the Technical University of Denmark during an invited stay there in 1973.

In the preparation of this book, the authors have benefited greatly from discussions with colleagues and former students. Special thanks are due to Bob Barnhill, Lothar Collatz, David Griffiths, Dirk Laurie, Jack Lambert, Peter Lancaster, Robin McLeod, Gil Strang, Gene Wachspress, Jim Watt and Olek Zienkiewicz. Final thanks are due to Ros Dudgeon and Doreen Manley for their expert typing of the manuscript.

Contents

Chapter 1

Introduction

1.1 APPROXIMATION BY PIECEWISE POLYNOMIALS

Consider initially the problem of approximating a real-valued function $f(x)$ over a finite interval of the x-axis. A simple approach is to break up the interval into a number of non-overlapping subintervals and to interpolate linearly between the values of $f(x)$ at the end points of each subinterval (see Figure 1(a)). If there are n subintervals denoted by $[x_i, x_{i+1}]$ $(i = 0,1,2, \ldots, n-1)$, then the piecewise linear approximating function depends only on the function values f_i $(= f(x_i))$ at the nodal points x_i $(i = 0,1,2, \ldots, n)$. In a problem where $f(x)$ is given implicitly by an equation (differential, integral, functional, etc.), the values f_i are the unknown parameters of the problem. In the problem of interpolation, the values f_i are known in advance.

In the subinterval $[x_i, x_{i+1}]$, the appropriate part of the linear

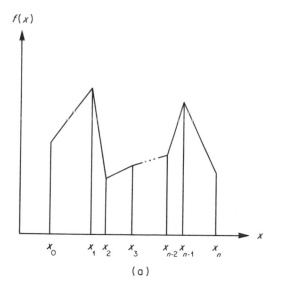

(a)

Figure 1a

approximating function is given by

$$p_1^{(i)}(x) = \alpha_i(x)f_i + \beta_{i+1}(x)f_{i+1} \quad (x_i \leqslant x \leqslant x_{i+1}), \tag{1.1}$$

where

$$\alpha_i(x) = \frac{x_{i+1} - x}{x_{i+1} - x_i} \quad \text{and} \quad \beta_{i+1}(x) = \frac{x - x_i}{x_{i+1} - x_i} \quad (i = 0,1,2, \ldots, n-1).$$

Hence the piecewise approximating function over the interval $x_0 \leqslant x \leqslant x_n$ is given by

$$p_1(x) = \sum_{i=0}^{n} \varphi_i(x)f_i, \tag{1.2}$$

where

$$\varphi_0(x) = \begin{cases} \dfrac{x_1 - x}{x_1 - x_0} & (x_0 \leqslant x \leqslant x_1) \\ 0 & (x_1 \leqslant x \leqslant x_n), \end{cases}$$

$$\varphi_i(x) = \begin{cases} 0 & (x_0 \leqslant x \leqslant x_{i-1}) \\ \dfrac{x - x_{i-1}}{x_i - x_{i-1}} & (x_{i-1} \leqslant x \leqslant x_i) \\ \dfrac{x_{i+1} - x}{x_{i+1} - x_i} & (x_i \leqslant x \leqslant x_{i+1}) \\ 0 & (x_{i+1} \leqslant x \leqslant x_n), \end{cases} \tag{1.3}$$

and

$$\varphi_n(x) = \begin{cases} 0 & (x_0 \leqslant x \leqslant x_{n-1}) \\ \dfrac{x - x_{n-1}}{x_n - x_{n-1}} & (x_{n-1} \leqslant x \leqslant x_n) \end{cases}$$

are pyramid functions illustrated in Figure 1(b). The pyramid functions given by (1.3) represent an elementary type of basis function. In particular the basis functions $\varphi_i(x)$ $(i = 1,2, \ldots, n-1)$ are identically zero except for the range $x_{i-1} \leqslant x \leqslant x_{i+1}$, and are said to have local support. Throughout this text, basis functions will be constructed of varying degrees of complexity but always with local support. A fundamental property of most basis functions is that they take the value unity at a particular nodal point and are zero at most of the other nodal points.

In general, the first derivatives of the piecewise approximating polynomial $p_1(x)$ given by (1.1) are not the same as $f(x)$ even at the nodes. Consequently we now look at the possibility of constructing a piecewise

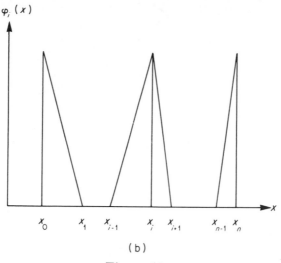

(b)

Figure 1b

approximating function which has the same values of function and first
derivative as $f(x)$ at the nodal points x_i ($i = 0,1,2, \ldots , n$). In mathematical terms, we have to construct a piecewise cubic polynomial $p_3(x)$
such that

$$\mathbf{D}^k f(x_i) = \mathbf{D}^k p_3(x_i) \quad (k = 0,1; i = 0,1,2, \ldots , n),$$

where $\mathbf{D} = d/dx$. In the subinterval $[x_i, x_{i+1}]$, the appropriate part of
the approximating cubic polynomials is given by

$$p_3^{(i)}(x) = \alpha_i(x)f_i + \beta_{i+1}(x)f_{i+1} + \gamma_i(x)f_i' + \delta_{i+1}(x)f_{i+1}', \qquad (1.4)$$

where

$$\alpha_i(x) = \frac{(x_{i+1} - x)^2 \, [(x_{i+1} - x_i) + 2(x - x_i)]}{(x_{i+1} - x_i)^3},$$

$$\beta_{i+1}(x) = \frac{(x - x_i)^2 \, [(x_{i+1} - x_i) + 2(x_{i+1} - x)]}{(x_{i+1} - x_i)^3},$$

$$\gamma_i(x) = \frac{(x - x_i)(x_{i+1} - x)^2}{(x_{i+1} - x_i)^2} \qquad (1.5)$$

and

$$\delta_{i+1}(x) = \frac{(x - x_i)^2 (x - x_{i+1})}{(x_{i+1} - x_i)^2}$$

($i = 0,1,2, \ldots , n - 1$) and where $'$ denotes differentiation with respect
to x. The piecewise approximating function over the interval

4

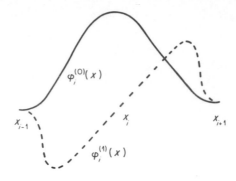

Figure 2

$x_0 \leqslant x \leqslant x_n$ is given by

$$p_3(x) = \sum_{i=0}^{n} [\varphi_i^{(0)}(x)f_i + \varphi_i^{(1)}(x)f_i'], \tag{1.6}$$

where the cubic polynomials $\varphi_i^{(0)}(x)$, $\varphi_i^{(1)}(x)$ ($i = 0,1,2, \ldots, n$) are easily obtained from (1.5). The basis functions $\varphi_i^{(0)}(x)$ and $\varphi_i^{(1)}(x)$ ($i = 1,2, \ldots, n-1$) are illustrated in Figure 2.

The basis functions in (1.2) and (1.6) arise from particular cases of *piecewise Hermite interpolation* (or approximation) for a partitioned interval. In more general terms, let $\Pi : a = x_0 < x_1 < \ldots < x_n = b$ denote any partition of the interval $R = [a, b]$ on the x-axis. For a positive integer m, and a partition Π of the interval, let $H = H^{(m)}(\Pi, R)$ be the set of all real-valued piecewise polynomial functions $w(x)$ defined on R such that $w(x) \in C^{m-1}(R)$ and $w(x)$ is a polynomial of degree $2m-1$ on each subinterval $[x_i, x_{i+1}]$ of R. Given any real-valued function $f(x) \in C^{m-1}(R)$, then its unique piecewise Hermite interpolate is the element $p_{2m-1}(x) \in H$ such that

$$D^k f(x_i) = D^k p_{2m-1}(x_i) \quad \begin{cases} (0 \leqslant k \leqslant m-1) \\ (0 \leqslant i \leqslant n). \end{cases} \tag{1.7}$$

The particular cases $m = 1,2$ have already been dealt with and produce the basis functions given by (1.2) and (1.6) respectively. Error estimates for piecewise Hermite interpolates are given by Birkhoff, Schultz and Varga (1968).

In problems where only $f(x)$ has to be determined, it is often undesirable to introduce derivatives of $f(x)$ as additional parameters and so cause a considerable increase in the order of the system of equations to be solved. Consequently a very desirable property in piecewise functions might be continuity of derivatives at the points at which pieces of the polynomials meet without introducing the values of the derivatives as additional unknown parameters. The simplest example of this approach is the fitting in each subinterval $[x_i, x_{i+1}]$ $(i = 0,1,2, \ldots, n-1)$ of a quadratic such that the first derivatives are continuous at each internal nodal point x_i $(i = 1,2, \ldots, n-1)$. A convenient form for this piece-wise approximate, known as the quadratic *spline* is

$$S_2^{(i)}(x) = f_i + \frac{f_{i+1} - f_i}{x_{i+1} - x_i} (x - x_i) + c_i(x - x_i)(x - x_{i+1})$$

$$(i = 0,1,2, \ldots, n-1) \tag{1.8}$$

and the continuity of the first derivatives leads to

$$c_i + c_{i-1} = \frac{1}{h^2} (f_{i+1} - 2f_i + f_{i-1}) \quad (i = 1,2, \ldots, n-1), \tag{1.9}$$

where the nodal points in the interval have been taken equally spaced, distance h apart. Equation (1.9) gives $(n-1)$ linear relations between the n unknown coefficients c_i $(i = 0,1,2, \ldots, n-1)$, and so in the case of the quadratic spline there is one free coefficient. Since $S_2^{(i)''}(x) = 2c_i$ $(i = 0,1,2, \ldots, n-1)$, a knowledge of the second derivative at any point completely solves the problem.

The most popular form of the spline is the cubic spline. Here, given the values of f_i $(i = 0,1,2, \ldots, n)$, we fit cubic polynomials between successive pairs of nodal points and require continuity of both first and second derivatives at all internal nodal points. In this case, if $S_3^{(i)}(x)$ $(i = 0,1,2, \ldots, n-1)$ is the required cubic spline, then $S_3^{(i)''}(x)$ must be linear in $[x_i, x_{i+1}]$, and so

$$S_3^{(i)''}(x) = c_i \frac{x_{i+1} - x}{x_{i+1} - x_i} + c_{i+1} \frac{x - x_i}{x_{i+1} - x_i} \quad (i = 0,1,2, \ldots, n-1)$$

where c_i, c_{i+1} are the values of the second derivatives at x_i, x_{i+1} respectively. This form ensures continuity of the second derivative at the internal nodal points. After applying the further conditions

$$\left. \begin{array}{l} S_3^{(i)}(x_i) = f_i \\ S_3^{(i)}(x_{i+1}) = f_{i+1} \end{array} \right\} \quad (i = 0,1,2, \ldots, n-1)$$

and

$$S_3^{(i-1)'}(x_i) = S_3^{(i)'}(x_i) \quad (i = 1,2, \ldots, n-1),$$

the cubic spline is obtained in the form

$$S_3^{(i)}(x) = \frac{c_i}{6h} (x_{i+1} - x)^3 + \frac{c_{i+1}}{6h} (x - x_i)^3 + \left(\frac{f_i}{h} - \frac{hc_i}{6} \right)(x_{i+1} - x)$$

$$+ \left(\frac{f_{i+1}}{h} - \frac{hc_{i+1}}{6} \right)(x - x_i) \quad (i = 0,1,2, \ldots, n-1), \tag{1.10}$$

where the nodal points are equally spaced, and the $(n + 1)$ coefficients c_i $(i = 0,1,2, \ldots, n)$ are given by the $(n - 1)$ linear relations

$$c_{i+1} + 4c_i + c_{i-1} = \frac{6}{h^2} (f_{i+1} - 2f_i + f_{i-1}) \quad (i = 1,2, \ldots, n-1).$$

(1.11)

The two free parameters in the case of the cubic spline are often removed by taking $c_0 = c_n = 0$, and hence the other parameters are uniquely defined by (1.11).

A more natural form of the cubic spline for equally spaced nodal points in the interval $I = [0, b]$ is

$$S_I \left(\frac{x}{h} \right) = \alpha_0 + \alpha_1 \left(\frac{x}{h} \right) + \alpha_2 \left(\frac{x}{h} \right)^2 + \alpha_3 \left(\frac{x}{h} \right)^3 + \sum_{s=1}^{n-1} \beta_s \left(\frac{x}{h} - s \right)_+^3 ,$$

(1.12)

where

$$\left(\frac{x}{h} - s \right)_+ = \begin{cases} 0 & \left(\frac{x}{h} \leqslant s \right) \\ \frac{x}{h} - s & \left(\frac{x}{h} > s \right) \end{cases} .$$

It can easily be verified that $S_I(x/h)$ and all its derivatives except the third are continuous at the $(n - 1)$ internal nodal points x_i $(i = 1,2, \ldots, n-1)$ for all values of the $(n + 3)$ coefficients $\alpha_0, \alpha_1, \alpha_2, \alpha_3, \beta_s$ $(s = 1,2, \ldots, n-1)$. Applying the condition

$$S_I \left(\frac{x_i}{h} \right) = f_i \quad (i = 0,1,2, \ldots, n), \qquad \bullet$$

(1.13)

there are $(n + 1)$ linear relations for the $(n + 3)$ coefficients, and so there are two free parameters. The system of linear equations reduces to the form given in Exercise 4. If the cubic spline (1.12) involving two arbitrary parameters is now expressed in the form

$$S_I \left(\frac{x}{h} \right) = \sum_{i=0}^{n} f_i C_i \left(\frac{x}{h} \right),$$

(1.14)

where $C_i(x_i/h) = 1$, $C_i(x_j/h) = 0$ $(j \neq i)$, $(i, j = 0,1,2, \ldots, n)$, the cardinal splines $C_i(x/h)$ obtained do not have local support and are not practical basis functions.

Cubic spline functions with local support of $4h$ were introduced as suitable basis functions by Schoenberg (1969). At nodal points $x = ih$ $(i = 2,3, \ldots, n-2)$, away from the ends of the interval, these take the form

$$B_i\left(\frac{x}{h}\right) = \frac{1}{4}\left[\left\{\frac{x}{h}-(i-2)\right\}_+^3 - 4\left\{\frac{x}{h}-(i-1)\right\}_+^3 + 6\left\{\frac{x}{h}-i\right\}_+^3\right.$$
$$\left. - 4\left\{\frac{x}{h}-(i+1)\right\}_+^3 + \left\{\frac{x}{h}-(i+2)\right\}_+^3\right]. \tag{1.15}$$

These functions and their first two derivatives are zero for $-\infty < x/h \leqslant i-2$ and $i+2 \leqslant x/h < +\infty$. Also

$$B_i(i-1) = B_i(i+1) = \frac{1}{4}, B_i(i) = 1 \quad (i = 2,3,\ldots,n-2).$$

The remaining functions $B_0(x/h)$, $B_1(x/h)$, $B_{n-1}(x/h)$, $B_n(x/h)$ require special consideration. By setting

$$S_I\left(\frac{x}{h}\right) = \sum_{i=0}^{n} \gamma_i B_i\left(\frac{x}{h}\right), \tag{1.16}$$

and matching the right-hand sides of (1.14) and (1.16), a tridiagonal system of linear equations is obtained which enables the coefficients γ_i ($i = 0,1,2,\ldots,n$) in (1.16) to be obtained. The majority of the equations in this system are given by

$$\gamma_i + \frac{1}{4}(\gamma_{i+1} + \gamma_{i-1}) = f_i \quad (i = 2,3,\ldots,n-2).$$

Bivariate approximation

We now consider the problem of approximating a real-valued function of two variables by piecewise continuous functions over a bounded region R with boundary ∂R. The region is divided up into a number of elements and the particular shapes of region considered at this stage are (1) rectangular and (2) polygonal.

(1) *Rectangular region.* The sides of this region are parallel to the x- and y-axes, and the region is subdivided into similar rectangular elements by drawing lines parallel to the axes. Let the rectangular region be $[x_0, x_m] \times [y_0, y_n]$ and a typical element be $[x_i, x_{i+1}] \times [y_j, y_{j+1}]$, where $x_{i+1} - x_i = h_1$ and $y_{j+1} - y_j = h_2$ ($0 \leqslant i \leqslant m-1, 0 \leqslant j \leqslant n-1$) (see Figure 3). The bilinear form which interpolates $f(x,y)$ over the rectangular element is

$$p_1^{(i,j)}(x,y) = \alpha_{i,j}(x,y)f_{i,j} + \beta_{i+1,j}(x,y)f_{i+1,j} + \gamma_{i,j+1}(x,y)f_{i,j+1}$$
$$+ \delta_{i+1,j+1}(x,y)f_{i+1,j+1}, \tag{1.17}$$

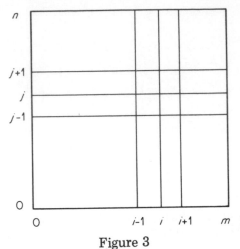

Figure 3

where

$$\alpha_{i,j}(x,y) = \frac{1}{h_1 h_2}(x_{i+1} - x)(y_{j+1} - y),$$

$$\beta_{i+1,j}(x,y) = \frac{1}{h_1 h_2}(x - x_i)(y_{j+1} - y),$$

$$\gamma_{i,j+1}(x,y) = \frac{1}{h_1 h_2}(x_{i+1} - x)(y - y_j)$$

and

$$\delta_{i+1,j+1}(x,y) = \frac{1}{h_1 h_2}(x - x_i)(y - y_j)$$

$(0 \leqslant i \leqslant m - 1; 0 \leqslant j \leqslant n - 1)$. The piecewise approximating function over the region $[x_0, x_m] \times [y_0, y_n]$ is given by

$$p_1(x,y) = \sum_{i=0}^{m} \sum_{j=0}^{n} \varphi_{i,j}(x,y)f_{i,j}. \tag{1.18}$$

The basis functions $\varphi_{i,j}(x,y)$ $(1 \leqslant i \leqslant m - 1; 1 \leqslant j \leqslant n - 1)$ are identically zero except for the rectangular region $[x_{i-1}, x_{i+1}] \times [y_{j-1}, y_{j+1}]$, and so have local support (see Exercise 6 and Figure 3).

The case just considered is the simplest example of piecewise *bivariate* Hermite interpolation (or approximation) over a rectangular region subdivided into rectangular elements. In more general terms, for any positive integer l, and any subdivision of the rectangle R into rectangular elements, let $H = H^{(l)}(R)$ be the collection of all real-valued

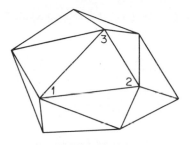

Figure 4

piecewise polynomials $g(x,y)$ defined on R such that $g(x,y) \in C^{l-1,\,l-1}(R)$ and $g(x,y)$ is a polynomial of degree $2l-1$ in each variable x and y on each rectangular element $[x_i, x_{i+1}] \times [y_j, y_{j+1}]$ $(0 \leqslant i \leqslant m-1;$ $0 \leqslant j \leqslant n-1)$ of R. Given any real-valued function $f(x,y) \in C^{l-1,\,l-1}(R)$, then its unique piecewise Hermite interpolant is the element $p_{2l-1}(x,y) \in H$ such that

$$\mathbf{D}^{(p,q)}f(x_i,y_j) = \mathbf{D}^{(p,q)}p_{2l-1}(x_i,y_j)$$

for all $0 \leqslant p,q \leqslant l-1$, $0 \leqslant i \leqslant m-1$, $0 \leqslant j \leqslant n-1$. The particular case $l = 1$ has already been dealt with and leads to bilinear basis functions of the type shown in Exercise 6. The case $l = 2$ is covered in Exercise 7. The interested reader is again referred to Birkhoff, Schultz and Varga (1968) for error estimates of bivariate Hermite interpolation.

(2) *Polygonal region*. This can either be a region in its own right or an approximation to a region of any shape. The polygon is subdivided in an arbitrary manner into triangular elements. In a typical triangular element with vertices (x_i,y_i) $(i = 1,2,3)$ (see Figure 4), the linear form which interpolates $f(x,y)$ over the triangular element is

$$p_1(x,y) = \sum_{i=1}^{3} \alpha_i(x,y)f_i, \tag{1.19}$$

where $f_i = f(x_i,y_i)$ $(i = 1,2,3)$. The coefficients $\alpha_i(x,y)$ $(i = 1,2,3)$ are given by

$$\alpha_1(x,y) = \frac{1}{C_{123}}(\tau_{23} + \eta_{23}x - \xi_{23}y),$$

$$\alpha_2(x,y) = \frac{1}{C_{123}}(\tau_{31} + \eta_{31}x - \xi_{31}y) \tag{1.20}$$

and

$$\alpha_3(x,y) = \frac{1}{C_{123}}(\tau_{12} + \eta_{12}x - \xi_{12}y),$$

where $|C_{123}|$ is twice the area of the triangle, and

$$\tau_{ij} = x_i y_j - x_j y_i,$$
$$\xi_{ij} = x_i - \dot{x}_j \quad (i,j = 1,2,3)$$

and

$$\eta_{ij} = y_i - y_j.$$

The functions given by (1.20) are of course only parts of the complete basis functions associated with vertices of a triangular network. The complete basis function with respect to any vertex is obtained by summing the appropriate parts associated with the triangles adjacent to the vertex. For example, the vertex 1 in Figure 4 has five adjacent triangles and so the basis function associated with this vertex has five parts. The complete basis function is known as a pyramid function.

Exercise 1 Show that the cubic polynomial $p_3(x)$ which takes the values

$$p_3(0) = f_0, \, p_3(1) = f_1, \, p_3'(0) = f_0', \, p_3'(1) = f_1',$$

is given by

$$p_3(x) = (1-x)^2(1+2x)f_0 + x(1-x)^2 f_0' + x^2(3-2x)f_1$$
$$+ x^2(x-1)f_1'.$$

Exercise 2 Use the result of Exercise 1 to obtain the coefficients in equation (1.5), and hence obtain the basis functions in equation (1.6).

Exercise 3 Using the method outlined in the text, obtain the equation of the cubic spline in the form (1.10), where the coefficients are given by (1.11).

Exercise 4 Applying the condition (1.13) to the spline given by (1.12), show that the system of equations for the coefficients in (1.12) reduces to

$$\beta_{i-1} + 4\beta_i + \beta_{i+1} = \delta^4 f_i \quad (i = 2,3,\ldots,n-2),$$
$$6\alpha_3 + 5\beta_1 + \beta_2 = \delta^3 f_{3/2},$$
$$2\alpha_2 + 6\alpha_3 + \beta_2 = \delta^2 f_1,$$
$$\alpha_1 + \alpha_2 + \alpha_3 = \delta f_{1/2},$$
$$\alpha_0 = f_0,$$

where δ is the usual central difference operator.

Exercise 5 Solve the set of equations in Exercise 4 for $\alpha_1 = \alpha_2 = 0$ and $n = 4$. Show that in this case, using (1.14), that

$$C_2\left(\frac{x}{h}\right) = \left(\frac{x}{h}-1\right)_+^3 - 8\left(\frac{x}{h}-2\right)_+^3 + 37\left(\frac{x}{h}-3\right)_+^3.$$

Sketch a rough graph of the cardinal spline $C_2(x/h)$ for $-\infty < x/h < +\infty$, and show that it does not have local support.

Exercise 6 In the unit square, show that the basis functions at internal nodes are given by

$$\varphi_{i,j}(x,y) = \begin{cases} \left[\dfrac{x}{h}-(i-1)\right]\left[\dfrac{y}{h}-(j-1)\right] \\ \qquad\qquad \left(i-1\leqslant\dfrac{x}{h}\leqslant i;\, j-1\leqslant\dfrac{y}{h}\leqslant j\right) \\[4pt] \left[\dfrac{x}{h}-(i-1)\right]\left[(j+1)-\dfrac{y}{h}\right] \\ \qquad\qquad \left(i-1\leqslant\dfrac{x}{h}\leqslant i;\, j\leqslant\dfrac{y}{h}\leqslant j+1\right) \\[4pt] \left[(i+1)-\dfrac{x}{h}\right]\left[\dfrac{y}{h}-(j-1)\right] \\ \qquad\qquad \left(i\leqslant\dfrac{x}{h}\leqslant i+1;\, j-1\leqslant\dfrac{y}{h}\leqslant j\right) \\[4pt] \left[(i+1)-\dfrac{x}{h}\right]\left[(j+1)-\dfrac{y}{h}\right] \\ \qquad\qquad \left(i\leqslant\dfrac{x}{h}\leqslant i+1;\, j\leqslant\dfrac{y}{h}\leqslant j+1\right) \\[4pt] 0 \qquad\qquad\qquad \text{elsewhere,} \end{cases}$$

where $1 \leqslant i,j \leqslant m-1$ and $mh = 1$.

Exercise 7 Consider the polynomial

$$g(x,y) = \sum_{r=0}^{3}\sum_{s=0}^{3}\alpha_{rs}x^r y^s$$

over the unit square $0 \leqslant x,y \leqslant 1$. Find the coefficients α_{rs} ($0 \leqslant r,s \leqslant 3$) in terms of the values of g, $\partial g/\partial x$, $\partial g/\partial y$ and $\partial^2 g/\partial x\partial y$ at the four corner points of the square. Show that the results of this calculation can be used to find the basis functions for the case $l = 2$ of the general theory of bivariate Hermite interpolation over a rectangular region subdivided into rectangular elements.

1.2 FUNCTION SPACES

This section contains an introduction to the mathematical structures required for an understanding of the theoretical developments of the finite element method. Only essential material will be presented and for a fuller treatment the interested reader is referred to Simmons (1963) or Yoshida (1965).

A *linear space* or *vector space* is a non-empty set X in which any two elements x and y, can be combined, by a process called *addition*, to give some element in X denoted by $x + y$, provided the process of addition satisfies the following conditions:

(i) $x + y = y + x$,

(ii) $x + (y + z) = (x + y) + z$,

(iii) there exists a zero element φ such that $\varphi + x = x + \varphi = x$ for all x,

(iv) for each x, there exists a negative $-x$ such that $x + (-x) = \varphi$.

It is also a necessary condition of a linear space that an element $x \in X$ can be combined with any real number or *scalar* α by *scalar multiplication* to give an element αx.

The process of scalar multiplication must satisfy the following conditions:

(v) $\alpha(x + y) = \alpha x + \alpha y$,

(vi) $(\alpha + \beta)x = \alpha x + \beta x$,

(vii) $(\alpha\beta)x = \alpha(\beta x)$,

(viii) $1.x = x$.

One example of a linear space is the set of all N-dimensional real vectors, where $\mathbf{a} + \mathbf{b} = \mathbf{c}$ is defined by $a_i + b_i = c_i$ $(i = 1, 2, \ldots, N)$ and $\alpha \mathbf{a} = \mathbf{d}$ by $d_j = \alpha a_j$ $(j = 1, 2, \ldots, N)$.

A normed linear space (n.l.s.) is a linear space on which there is defined a norm $\| x \|$ such that

(i) $\| x \| \geqslant 0$,

(ii) $\| x \| = 0 \Leftrightarrow x = 0$,

(iii) $\| x + y \| \leqslant \| x \| + \| y \|$,

(iv) $\| \alpha x \| = | \alpha | \, \| x \|$.

Thus we have the concept of the length of an element in the linear space. A *semi-norm* satisfies (i), (iii), and (iv) but not (ii).

An inner product space (i.p.s.) or scalar product space is a linear space in which there is defined a real-valued function (x,y) for each pair of vectors in the linear space, such that

(i) $(\alpha x + \beta y, z) = \alpha(x,z) + \beta(y,z)$,

(ii) $(x,y) = (y,x)$.

Exercise 8 Show that an i.p.s. is a n.l.s. with respect to the norm

$$\| x \| = (x,x)^{1/2}.$$

Then verify the parallelogram law

$$\| x + y \|^2 + \| x - y \|^2 = 2 \| x \|^2 + 2 \| y \|^2.$$

Show also that

$$4xy = \| x + y \|^2 - \| x - y \|^2.$$

Let $\{x_n\}$ be a sequence of points in an i.p.s., then

(a) $\{x_n\}$ is a *Cauchy sequence* if for each $\epsilon > 0$, there exists some $N = N(\epsilon)$ such that for all $n,m \geqslant N$,

$$\| x_n - x_m \| < \epsilon,$$

(b) $\{x_n\}$ is a *convergent sequence* if there exists a point x in the i.p.s. such that for each $\epsilon > 0$ there exists some $N = N(\epsilon)$ such that for all $n \geqslant N$,

$$\| x_n - x \| < \epsilon.$$

Exercise 9 Show that a convergent sequence is a Cauchy sequence.

To show the converse is untrue consider the following. Let X be the space of points on the interval $(0,1]$ and let $x_n = 1/n$. Then $\{x_n\}$ is a Cauchy sequence, but it is not a convergent sequence since the origin is not in the space.

A complete i.p.s. is one in which all Cauchy sequences are convergent sequences, and such a space is called a *Hilbert space*.

So far the spaces introduced have been such that a point in the space has represented a point on the real line, a vector or a matrix. In order to provide a Hilbert space which is readily applicable to the development of finite element methods, it is necessary to introduce a space in which the points represent functions. The most useful function spaces can be developed from a simple Hilbert space denoted by $\mathscr{L}_2(R)$, where R is, for simplicity, an interval $[a,b]$ of the real line. Functions $f(x)$ are points in this space if and only if

$$\int_a^b f^2(x)dx$$

is finite. Such functions are said to be *measurable*. For any two points $u(x)$ and $v(x)$, the inner product is defined by

$$(u,v) = \int_a^b u(x)v(x)dx,$$

and the norm by

$$\| u \|^2 = \int_a^b u^2(x)\mathrm{d}x.$$

Addition is defined by $(u + v)(x) = u(x) + v(x)$.

Exercise 10 Show that if $\| u \|$ and $\| v \|$ are finite, then (u,v) is also finite.

A subset of a linear space which is itself a linear space is called a *subspace*.

Exercise 11 Let $\mathscr{L}_2(R)$ be as previously defined. Which of the following subsets are subspaces?

   (i)   The functions u such that $u(a) = 0$.
   (ii) The functions u such that $(\mathrm{d}u/\mathrm{d}x)_{x=a} = 0$ and $u(b) = 0$.

Exercise 12 Let K be a proper subspace of the linear space $\mathscr{H}$, and let X be the set $\{x + h\}$, where x is some fixed point in $\mathscr{H}$ but not in K, and h is any point in K. Show that X, denoted by $\{x\} \oplus K$ is not a linear space.

The set X is called a *linear manifold*.

Let T be a mapping of the Hilbert space $\mathscr{H}$ onto itself. T is a linear operator if and only if

   (i) $T(x + y) = T(x) + T(y)$ (for all x, $y \in \mathscr{H}$),
   (ii) $T(\alpha x) = \alpha T(x)$ (for any scalar α).

A linear operator T is said to be *bounded* if there exists a constant $M > 0$ such that

$$\| Tx \| \leqslant M \| x \| \quad \text{(for all } x \in \mathscr{H})$$

and the smallest such M is called the norm of T and denoted by $\| T \|$; it follows that

$$\| T \| = \sup_{x \neq 0} \left\{ \frac{\| Tx \|}{\| x \|} \right\} = \sup_{\substack{x \neq 0 \\ \|x\| \leqslant 1}} \left\{ \frac{\| Tx \|}{\| x \|} \right\} = \sup_{\|x\|=1} \{ \| Tx \| \}.$$

A bounded linear operator is *continuous*; for if the sequence $\{x_n\}$ has the limit point x then the sequence $\{Tx_n\}$ has the limit point Tx.

Let F be a linear mapping of $\mathscr{H}_1 \times \mathscr{H}_2$ into $\mathscr{H}_3$, that is, for any $x_1 \in \mathscr{H}_1$, $x_2 \in \mathscr{H}_2$, $F(x_1, x_2) \in \mathscr{H}_3$ and F is linear in each of the arguments, then F is *bounded* if there exists $M > 0$ such that

$$\| F(x_1, x_2) \|_{\mathscr{H}_3} \leqslant M \| x_1 \|_{\mathscr{H}_1} \| x_2 \|_{\mathscr{H}_2}$$

and $\| F \|$ is the smallest such M. We shall denote by $\mathscr{L}(\mathscr{H}_1; \mathscr{H}_2)$ the space of bounded linear mappings from $\mathscr{H}_1$ into $\mathscr{H}_2$. The space

$\mathscr{L}(\mathscr{H};\mathbb{R})$ of *bounded linear functionals* is called the *dual* of $\mathscr{H}$ and is denoted by $\mathscr{H}'$. Elements of $\mathscr{L}(\mathscr{H}\times\mathscr{H};\mathbb{R})$ are referred to as *bilinear forms* (on $\mathscr{H}$).

Exercise 13 Let $E\in\mathscr{L}(\mathscr{H}_1\times\mathscr{H}_2;\mathscr{H}_3)$ and define $F\in\mathscr{L}(\mathscr{H}_1;\mathscr{H}_3)$ such that for some fixed $v\in\mathscr{H}_2$

$$F(u)=E(u,v)\quad\text{(for all }u\in\mathscr{H}_1).$$

Prove that

$$\|F\|\leqslant\|E\|\,\|v\|_{\mathscr{H}_2}.$$

Note that throughout this book subscripts will be frequently omitted from operator norms in the interest of brevity.

Theorem 1.1 Riesz Representation Theorem (Yosida, 1965, p. 90)
$F\in\mathscr{H}'$ *if and only if there exists a unique* $v\in\mathscr{H}$ *such that for all* $u\in\mathscr{H}$,

$$F(u)=(u,v)$$

and

$$\|F\|_{\mathscr{H}'}=\|v\|_{\mathscr{H}}.$$

There is a 1—1 correspondence between $\mathscr{H}$ and $\mathscr{H}'$ defined by Theorem 1.1; such a correspondence is called an *isomorphism* and as the two spaces have the same structure they are said to be *isomorphic*. From the definition of an operator norm it follows that

$$\|F\|_{\mathscr{H}'}=\sup_{\|u\|\neq0}\left\{\frac{|F(u)|}{\|u\|_{\mathscr{H}}}\right\}$$

and from Theorem 1.1 it follows that there exists $v\in\mathscr{H}$ such that

$$F(u)=(u,v).$$

As it is possible to identify each element of $\mathscr{H}'$ with a unique element of $\mathscr{H}$ there is no ambiguity if we also write

$$\|v\|_{\mathscr{H}'}=\sup_{\|u\|\neq0}\left\{\frac{|(u,v)|}{\|u\|}\right\}.$$

A linear operator T which maps the whole of a Hilbert space $\mathscr{H}$ onto a particular subspace K is called a *projection* if and only if it maps the points of K onto themselves, i.e.

$$Ty=y\quad\text{(for all }y\in K).$$

Exercise 14 Which of the following linear operators are projections?

(i) T which maps the 2-vector $\begin{pmatrix}\alpha\\\beta\end{pmatrix}$ into the vector $\begin{pmatrix}\alpha\\0\end{pmatrix}$.

16

(ii) Q which maps the space $\mathscr{L}_2(R)$, $R = [a, b]$ onto the space of linear functions, by linear interpolation at the end-points.

(iii) S which maps the space of 2 x 2 matrices into diagonal matrices by

$$S\left(\begin{bmatrix} \alpha & \beta \\ \gamma & \delta \end{bmatrix}\right) = \begin{bmatrix} \alpha + \delta & \\ & \beta + \gamma \end{bmatrix}$$

(iv) S^2, with S as in (iii).

A projection P, is said to be an *orthogonal projection* if, for all x in the space, and all y in the subspace,

$$(x - Px, y) = 0 \quad \text{(denoted by } (x - Px) \perp y),$$

i.e. the remainder $x - Px$ is orthogonal to all y in the subspace K. The remainder is said to belong to the orthogonal complement of K, which is denoted by $K^\perp$.

Lemma 1.1   *If $\mathscr{H}$ is a Hilbert space and K is any subspace of $\mathscr{H}$, then $K^\perp$ is also a Hilbert space.*

Lemma 1.2   *If P is an orthogonal projection of a Hilbert space $\mathscr{H}$, onto some subspace K, then $I - P$ is an orthogonal projection onto $K^\perp$ and for any $x \in \mathscr{H}$, there exists $u \in K$ and $v \in K^\perp$ such that $x = u + v$, that is, $\mathscr{H} = K \oplus K^\perp$.*

Exercise 15   Prove Lemma 1.1 and Lemma 1.2.

Theorem 1.2   *The orthogonal projection P of a Hilbert space onto a subspace is unique, and the 'length' of the remainder $\| x - Px \|$ is the minimum distance from x to the subspace.*

Proof   Assume that the orthogonal projection P, is not unique. Then there exists another projection Q, such that

$$(x - Qx, y) = 0 \quad \text{(for all } y \in K).$$

Thus, since $Px - Qx \in K$,

$$0 = (x - Qx, Px - Qx)$$
$$= (x - Px, Px - Qx) + (Px - Qx, Px - Qx)$$
$$= 0 + \| Px - Qx \|^2 > 0$$

since $Px \neq Qx$, and so there is a contradiction. The original assumption is thus false, and so the orthogonal projection is unique.

Now let v be any point in K, other than Px, then as above

$$(v - Px) \perp (x - Px).$$

Hence

$$\| x - v \|^2 = \| (x - Px) + (Px - v) \|^2$$
$$= \| x - Px \|^2 + 2(x - Px, Px - v) + \| Px - v \|^2$$
$$= \| x - Px \|^2 + \| Px - v \|^2 > \| x - Px \|^2,$$

which is the desired result.

Corollary (i) $\| Px \|^2 = (x, Px)$ and $\| x - Px \|^2 = \| x \|^2 - (x, Px)$.

Corollary (ii)   *For any $x \in \mathscr{H}$ there exists a unique $u_0 \in K$ and a unique $v_0 \in K^\perp$ such that, $x = u_0 + v_0$ and*

$$\underset{u \in K}{\text{minimum}} \| x - u \|^2 = \| x - u_0 \|^2 = \| v_0 \|^2. \tag{1.21}$$

It is possible to replace K by $K^\perp$ in (1.21) to give a different minimum principle;

$$\underset{v \in K^\perp}{\text{minimum}} \| x - v \|^2 = \| u_0 \|^2.$$

Since

$$\| x \|^2 = \| u_0 \|^2 + \| v_0 \|^2,$$

we can replace this minimum principle by a *maximum* principle

$$\underset{v \subset K}{\text{maximum}} \{ \| x \|^2 - \| x - v \|^2 \}$$

or

$$\underset{v \in K}{\text{maximum}} \{ 2(x, v) - \| v \|^2 \}, \tag{1.22}$$

which has the same solution as (1.21).

A linear operator T, is said to be *positive definite* if and only if

$$(Tx, x) > 0 \quad \text{(for all } x \neq 0 \text{),}^\dagger$$

or positive semi-definite if

$$(Tx, x) \geqslant 0 \quad \text{(for all } x \neq 0 \text{).}$$

For example, the mapping $\begin{pmatrix} \alpha \\ \beta \end{pmatrix}$ to $\begin{pmatrix} \alpha/2 \\ \beta/2 \end{pmatrix}$ is positive definite, whereas

the mapping $\begin{pmatrix} \alpha \\ \beta \end{pmatrix}$ to $\begin{pmatrix} \alpha \\ 0 \end{pmatrix}$ is positive semi-definite, since

$$(Tx, x) = \alpha^2$$

which attains the value zero whenever $\alpha = 0$, for arbitrary β.

---

†Some authors use the definition $(Tx, x) \geqslant \gamma \| x \|^2$. $(\gamma > 0)$.

The *adjoint* of T is the operator T^* such that

$$(Tx, y) = (x, T^*y) \quad \text{(for all } x, y\text{).}$$

If $(Tx, y) = (x, Ty)$, then T is *self-adjoint*.

1.3 APPROXIMATING SUBSPACES

Consider a Hilbert space $\mathscr{H}$ and let $S = \{x_1, \ldots, x_N\}$ be a set of N elements of $\mathscr{H}$. It is assumed that S is a linearly independent set, i.e. there is no set of scalars α_i ($i = 1, 2, \ldots, N$), excluding the zero set $\alpha_1 = \alpha_2 = \cdots = \alpha_N = 0$, such that

$$\alpha_1 x_1 + \alpha_2 x_2 + \cdots + \alpha_N x_N = 0.$$

Then if for every element $x \in \mathscr{H}$, there exist scalars β_i ($i = 1, 2, \ldots, N$) such that

$$x = \beta_1 x_1 + \beta_2 x_2 + \cdots + \beta_N x_N,$$

the set S is said to form a *basis* for $\mathscr{H}$, which is said to be an N-*dimensional space*.

For example, the space of vectors of the form $\begin{pmatrix} \alpha \\ \beta \end{pmatrix}$ is a two-dimensional space. $S_1 = \left\{ e_1 = \begin{pmatrix} 0 \\ 1 \end{pmatrix}, e_2 = \begin{pmatrix} 1 \\ 0 \end{pmatrix} \right\}$ forms a basis as does $S_2 = \left\{ f_1 = \frac{1}{\sqrt{2}} \begin{pmatrix} 1 \\ 1 \end{pmatrix}, f_2 = \frac{1}{\sqrt{2}} \begin{pmatrix} 1 \\ -1 \end{pmatrix} \right\}$. These two bases are illustrated in Figure 5.

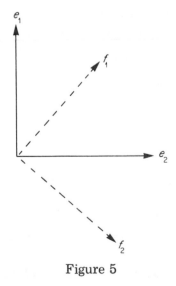

Figure 5

Exercise 16   Prove that any two bases of the same space must have the same number of elements.

The importance of finite-dimensional function spaces becomes apparent when the approximating functions mentioned in Section 1.1 are considered as elements of $\mathscr{L}_2(R)$. For example, if the interval $R = [a, b]$ is partitioned by the points x_i ($i = 0,1, \ldots, n$) it is possible to construct the set of Hermite functions which are piecewise linear on the interval. Any linearly independent set of $(n + 1)$ such functions constitutes a basis for $H^{(1)}(\Pi, R)$. It is a simple matter to show that H is a complete subspace of $\mathscr{L}_2(R)$, the details are left as an exercise for the reader, and so H is an $(n + 1)$ dimensional subspace of $\mathscr{L}_2(R)$. In fact the pyramid functions given by (1.3) constitute a natural choice of basis for H in view of the significance of the parameters

$$f_i \ (i = 0,1, \ldots, n).$$

Exercise 17   Let the partition Π be given by $x_i = i/4$ ($i = 0, \ldots, 4$) and let the functions $D_i(x)$ be defined by

$$D_i(x) = \sum_{k=0}^{4} \alpha_k^{(i)} \varphi_k \qquad (i = 0, \ldots, 4),$$

where if $\alpha^{(i)} = (\alpha_0^{(i)}, \ldots, \alpha_4^{(i)})$, then $\alpha^{(0)} = (1,1,1,1,1)$, $\alpha^{(1)} = (1,\sqrt{\tfrac{1}{2}},0,-\sqrt{\tfrac{1}{2}},-1)$, $\alpha^{(2)} = (1,0,-1,0,1)$, $\alpha^{(3)} = (1,-\sqrt{\tfrac{1}{2}},0,\sqrt{\tfrac{1}{2}},-1)$ and $\alpha^{(4)} = (1,-1,1,-1,1)$. Figure 6 illustrates the function $D_3(x)$ together with $\cos(3\pi x)$ in $[0,1]$.

(i)   Sketch the graphs of $D_i(x)$ ($i = 0, \ldots, 4$).
(ii) Evaluate the scalar products $(D_i(x), D_j(x))$, and then comment on the significance of the $D_i(x)$ as a basis for Π and of the coefficients g_i in the function

$$g(x) = \sum_{i=0}^{4} g_i D_i(x).$$

Exercise 18   Give a basis for the space of cubic spline functions and the space of piecewise planar functions defined on a triangular mesh.

The approximating functions mentioned in Section 1.1 are all particular cases of a general approximation problem. This problem is to associate with every element f, of a Hilbert space $\mathscr{H}$, a unique element $\tilde{f}$, of an N-dimensional approximating subspace K (for example, if $K = H$ then $N = n + 1$). Then $\tilde{f}$ is said to be the K-approximate of f. The mapping of f into $\tilde{f}$ is usually a linear mapping and is invariably a projection. For example if $K = H$, $\tilde{f}$ can be found by linear interpolation between the partition points. This is a unique mapping which is also a projection.

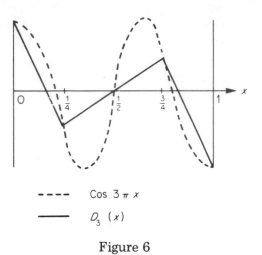

$$- - - - \quad \text{Cos } 3 \pi x$$

$$\underline{\hspace{1cm}} \quad D_3 (x)$$

Figure 6

Exercise 19   Construct a mapping from $\mathscr{L}_2(R)$ into H which is not a projection.

Once an approximation to any $f \in \mathscr{L}_2(R)$ has been constructed, it is reasonable to ask the question 'How good is the approximation?'. The approximation is *best* if the element $\tilde{f} \in K$ is such that the length of the error $f - \tilde{f}$ is minimized. In Theorem 1.2 it was shown that if P is the orthogonal projection onto K, then $\| f - Pf \|$ is the minimum distance from f to the subspace K, and hence the best approximation is $\tilde{f} = Pf$. Since it is assumed that K is of finite dimension, this can be used to calculate the best approximate. Assume that the set $S = \{f_1, \ldots, f_N\}$ forms a basis for K, then since

$$(f - \tilde{f}, g) = 0$$

for all $g \in K$, it follows that

$$\left(f - \tilde{f}, \sum_{\beta=1}^{N} \beta_k f_k \right) = 0$$

for all possible sets of scalars β_k $(k = 1, \ldots, N)$, and so

$$(f - \tilde{f}, f_k) = 0 \quad (k = 1, \ldots, N). \tag{1.23}$$

If $\tilde{f} = \sum_{i=1}^{N} \alpha_i f_i$, it follows that

$$(f, f_k) - \sum_{i=1}^{N} \alpha_i (f_i, f_k) = 0 \quad (k = 1, \ldots, N).$$

This can be written as

$$G\alpha = b,$$ \hfill (1.24)

where

$$G = [g_{ik}] = [(f_i, f_k)],$$
$$\alpha = \{\alpha_1, \ldots, \alpha_N\}^T$$

and

$$b = \{(f, f_1), \ldots, (f, f_N)\}^T.$$

The matrix G is called the Gram matrix, and the equations (1.24) are called *normal equations*.

Exercise 20 Verify that for $\mathcal{H} = \mathcal{L}_2(R)$, with the inner product and norm given by (1.21) and (1.22) respectively, the best approximate $\tilde{f}$ as defined by (1.23) is also best in the least squares sense. Note that the normal equations (1.24) are not recommended for calculating the solution to a least squares problem since they tend to be ill conditioned.

Exercise 21 Verify that it is possible to define the Hilbert space $\mathcal{L}_2^w(R)$ using the inner product

$$(u, v)_w = \int_a^b w(x)u(x)v(x)dx,$$

where $w(x) \geq 0$ is a suitable weight function. Hence calculate the normal equations which determine the best weighted least squares approximation of a function $f(x)$ in $\mathcal{L}_2^w(R)$.

Exercise 22 For functions $u(x)$ and $v(x)$ that are measurable and have a measurable first derivative, verify that it is possible to define the Hilbert space $\mathcal{H}_2^{(1)}(R)$ using the inner product

$$(u, v)_1 = \int_a^b \{u(x)v(x) + u'(x)v'(x)\}dx$$

and the norm

$$\|u\|_1^2 = \int_a^b \{u^2(x) + u'^2(x)\}dx$$

where dash indicates differentiation with respect to x. Derive the normal equations for best approximation of a function $f(x)$ in this space. The space $\mathcal{H}_2^{(1)}(R)$ is an example of a *Sobolev* space.

Methods of approximation involving the solution of a set of normal equations to obtain the orthogonal projection of a function onto a finite dimensional subspace are referred to as *projection* methods of approximation.

Chapter 2

Variational Principles

2.1 INTRODUCTION

Variational principles occur widely in physical and other problems and approximate methods of solution of such problems are often based on associated variational principles. The mathematical formulation of a variational principle is that the integral of some typical function has a smaller (or larger) value for the actual performance of the system than for any virtual performance subject to the general conditions of the system. The integrand is a function of coordinates, field amplitudes and their derivatives, and the integration is over a region governed by the coordinates of the system, which may include the time. The problem of determining the minimum of the integral often leads to one or more partial differential equations together with appropriate boundary conditions. It is not the intention of this book to give approximate methods of solution of these differential equations as a means of solving the original physical problems formulated as variational principles. Instead we intend to outline an approximate method which is directly based on the variational principle.

As an example of determining the extremum of an integral, consider the double integral

$$I(u) = \iint_R F(x, y, u, u_x, u_y) dx dy, \tag{2.1}$$

where u is continuous and has continuous derivatives up to the second order, and takes prescribed values on the boundary of R, which is a finite region in the (x,y) plane. It is relatively easy (Courant and Hilbert, 1953) to see that the necessary condition for $I(u)$ to have an extremum is that $u(x,y)$ must satisfy the *Euler–Lagrange* differential equation

$$\frac{\partial}{\partial x} F_{u_x} + \frac{\partial}{\partial y} F_{u_y} - F_u = 0. \tag{2.2}$$

From the many solutions of this equation, the particular solution is selected which satisfies the given boundary conditions. For example, when

$$F = \tfrac{1}{2}(u_x^2 + u_y^2),$$

(2.2) reduces to Laplace's equation,

$$\frac{\partial^2 u}{\partial x^2} + \frac{\partial^2 u}{\partial y^2} = 0.$$

The reason for demanding continuous second derivatives for u is now clear; it is to justify the existence of the Euler equation. Approximate methods based on the minimization of I in (2.1), however, only require continuity of u and piecewise continuity of the first derivatives.

We now return to the basic problem of the variational principle which is to determine the function from an admissible class of functions such that a certain definite integral involving the function and some of its derivatives takes on a maximum or minimum value in a closed region R. This is a generalization of the elementary theory of maxima and minima of the calculus which is concerned with the problem of finding a point in a closed region at which a function has a maximum or minimum value compared with neighbouring points in the region. The definite integral in the variational principle is an example of a *functional* and it depends on the entire course of a function rather than on a number of variables. The domain of the functional is the space of the admissible functions. The main dificulty with the variational principle approach is that problems which can be meaningfully formulated as variational principles may not have solutions. This is reflected in mathematical terms by the domain of admissible functions of the functional not forming a closed set. *Thus the existence of an extremum (maximum or minimum) cannot be assumed for a variational principle.* However, in this text we are concerned with *approximate* solutions of variational principles. These are obtained by considering some closed subset of the space of admissible functions to provide an upper and lower bound for the theoretical solution of the variational principle.

One apparent advantage of the variational approach is that we seem to require less continuity in the solution function. This paradox is explained at length on pages 199—204 of Courant and Hilbert (1953) and Chapter 2 of Clegg (1967). As a consequence of the weaker continuity requirements, a useful advantage of the variational approach is the greater ease with which approximate solutions can be constructed. A large part of the present text is devoted to describing such approximate methods.

The space over which the integral is evaluated in a variational principle may contain the time coordinate. We shall look first at variational principles which do not involve the time. These variational principles, usually of minimum potential energy type, govern problems of stable equilibrium which arise from classical field problems of

24

mathematical physics. This chapter contains only the material on variational principles which is relevant to the main theme of this book. No proofs or detailed discussions are given and the interested reader is referred to appropriate parts of books such as Courant and Hilbert (1953), Morse and Feshbach (1953), Hildebrand (1965), Schechter (1967) and Clegg (1967).

2.2 STABLE EQUILIBRIUM PROBLEMS

The differential equation which is associated with a variational principle is known as the *Euler—Lagrange* equation. It is a necessary, but rarely sufficient, condition which a function must satisfy if it is to maximize or minimize a definite integral. The simplest problem of the variational calculus is to determine the minimum of the integral

$$I(u) = \int_{x_0}^{x_1} F(x,u(x),u'(x))dx,$$

where the values $u(x_0)$ and $u(x_1)$ are given, and a dash denotes differentiation with respect to x. The necessary but not sufficient condition for the minimum to exist is that $u(x)$ satisfies the differential equation

$$\frac{\partial F}{\partial u} - \frac{d}{dx}\frac{\partial F}{\partial u'} = 0.$$

We shall now generalize this result to cover the following cases:

(1) *Two dependent variables.*   The integral to be minimized is

$$I(u,v) = \int_{x_0}^{x_1} F(x,u(x),v(x),u'(x),v'(x))dx,$$

where the values $u(x_0)$, $u(x_1)$, $v(x_0)$, $v(x_1)$ are given. The necessary conditions are

$$\frac{\partial F}{\partial u} - \frac{d}{dx}\frac{\partial F}{\partial u'} = 0,$$

$$\frac{\partial F}{\partial v} - \frac{d}{dx}\frac{\partial F}{\partial v'} = 0.$$

(2) *Two independent variables.*   The integral is

$$I(u) = \iint_R F(x,y,u(x,y),u_x(x,y),u_y(x,y))dx\,dy,$$

where u takes on prescribed values on the boundary of R, the region of integration. The necessary condition is

$$\frac{\partial F}{\partial u} - \frac{\partial}{\partial x}\frac{\partial F}{\partial u_x} - \frac{\partial}{\partial y}\frac{\partial F}{\partial u_y} = 0.$$

(3) *Higher derivatives.*   For variational principles involving second derivatives, the integral is

$$I(u) = \int_{x_0}^{x_1} F(x,u(x),u'(x),u''(x))dx,$$

where the values $u(x_0)$, $u'(x_0)$, $u(x_1)$, $u'(x_1)$ are given and the corresponding necessary condition is

$$\frac{\partial F}{\partial u} - \frac{d}{dx}\frac{\partial F}{\partial u'} + \frac{d^2}{dx^2}\frac{\partial F}{\partial u''} = 0.$$

(4) *Constrained extrema.*   Here the variational problem is constrained by one or more auxiliary conditions. One integral expression is to be made an extremum (maximum or minimum) while one or more other integral expressions maintain fixed values. Such problems are termed *isoperimetric* problems. Consider for example the problem of determining the function $u(x)$ which maximizes (or minimizes) the integral

$$I(u) = \int_{x_0}^{x_1} F(x,u(x),u'(x))dx,$$

subject to the condition that $u(x)$ satisfies the equation

$$\int_{x_0}^{x_1} G(x,u(x),u'(x))dx = \alpha, \qquad (2.3)$$

where the constant α is given. The necessary condition for an extremum to exist is that

$$\frac{\partial(F + \lambda G)}{\partial u} - \frac{d}{dx}\frac{\partial(F + \lambda G)}{\partial u'} = 0,$$

where the numerical value of the parameter λ is chosen so that (2.3) is satisfied. A simple example of an isoperimetric problem is the catenary. The problem is to find the shape of a uniform heavy string with fixed end points which hangs under gravity. Here we require to find the function $u(x)$ which passes through the points (x_0, u_0) and (x_1, u_1) and makes the integral

$$\int_{x_0}^{x_1} u(1 + u'^2)^{1/2}\, dx$$

as small as possible, while maintaining a fixed value for the integral

$$\int_{x_0}^{x_1} (1 + u'^2)^{1/2}\,dx.$$

Exercise 1   Show that the length of the curve connecting two points (x_0, u_0) and (x_1, u_1) is

$$I(u) = \int_{x_0}^{x_1} (1 + u'^2)^{1/2}\,dx.$$

Use the associated Euler—Lagrange equation to find the shortest path between the two points.

Exercise 2   Find the function $u(x)$ which passes through the points (x_0, u_0) and (x_1, u_1) and gives the minimum surface of revolution when rotated about the x-axis.

Exercise 3   Show that

$$\frac{\partial^2 u}{\partial x^2} + \frac{\partial^2 u}{\partial y^2} + \frac{\partial^2 u}{\partial z^2} + f(x,y,z) = 0$$

is the necessary condition for the integral

$$I(u) = \iiint_R \tfrac{1}{2}[u_x^2 + u_y^2 + u_z^2 - 2u\,f(x,y,z)]\,dxdydz$$

to be a minimum when u is specified on the surface ∂R which surrounds the volume R.

Before proceeding further, we give some examples of variational principles and equivalent Euler—Lagrange equations. In these examples the region under consideration is R with ∂R as its boundary.

(1) *Dirichlet problem for Laplace's equation.*

$$I(u) = \iint_R \tfrac{1}{2}(u_x^2 + u_y^2)\,dxdy \quad (u \text{ given on } \partial R),$$

$$u_{xx} + u_{yy} = 0.$$

(2) *Loaded and clamped plate.*   (Biharmonic operator.)

$$I(u) = \iint_R \tfrac{1}{2}[u_{xx}^2 + 2u_{xy}^2 + u_{yy}^2 - 2qu]\,dxdy$$

$$\left(u = \frac{\partial u}{\partial n} = 0 \text{ on } \partial R\right),$$

$$u_{xxxx} + 2y_{xxyy} + u_{yyyy} = q(x,y),$$

$q(x,y)$ is the normal load on the plate.

(3) *Small displacement theory of elasticity.* (Plane stress.)

$$I(u,v) = \iint_R \tfrac{1}{2}[(1-v)(u_x^2 + v_y^2) + v(u_x + v_y)^2$$
$$+ \tfrac{1}{2}(1-v)(u_y + v_x)^2]\,dxdy \quad (u,v \text{ given on } \partial R),$$
$$u_{xx} + vv_{xy} + \tfrac{1}{2}(1-v)(u_{yy} + v_{xy}) = 0,$$
$$v_{yy} + vu_{xy} + \tfrac{1}{2}(1-v)(u_{xy} + v_{xx}) = 0.$$

(4) *Radiation (e^u) and molecular diffusion (u^2).*

$$I(u) = \iint_R \tfrac{1}{2}\left(u_x^2 + u_y^2 + \left\{\begin{array}{l} 2e^u \\ \tfrac{2}{3}u^3 \end{array}\right.\right)dxdy \quad (u \text{ given on } \partial R),$$

$$u_{xx} + u_{yy} = \left\{\begin{array}{l} e^u \\ u^2. \end{array}\right.$$

(5) *Plateau's problem.* (To find the surface of minimum area bounded by a closed curve in three-dimensional space.)

$$I(u) = \iint_R \tfrac{1}{2}(1 + u_x^2 + u_y^2)^{1/2}\,dxdy \quad (u \text{ given on } \partial R),$$
$$\nabla(\gamma_1)\nabla u = 0, \quad \gamma_1 = (1 + u_x^2 + u_y^2)^{-1/2}.$$

(6) *Non-Newtonian fluids.*

$$I(u) = \iint_R [\tfrac{1}{2}(u_x^2 + u_y^2)^{1+s} + uc]\,dxdy \quad (u \text{ given on } \partial R),$$
$$\nabla(\gamma_2)\nabla u = c \text{ (constant)}, \quad \gamma_2 = (u_x^2 + u_y^2)^s \quad (0 \geqslant s \geqslant -\tfrac{1}{2}).$$

(7) *Compressible flow.*

$$I(p) = \iint_R p\,dx \qquad \begin{array}{l} \rho \text{ density,} \\ p \text{ pressure,} \\ \varphi \text{ velocity potential.} \end{array}$$
$$(\rho\,\nabla\,\varphi) = 0$$

Of the problems above, the first three are linear, the fourth is mildly non-linear, and the last three are non-linear.

2.3 BOUNDARY CONDITIONS

The steady problems discussed in the previous section have been such that the function is specified on the boundary and so is not subject to variation there. In many problems, however, the function is not specified on the boundary and alternative equally valid boundary conditions apply. Consider, for example (Courant and Hilbert, 1953, pp. 208, 209), the variational problem consisting of the minimization of the integral

$$I(u) = \int_{x_0}^{x_1} F(x,u,u')dx, \qquad (2.4)$$

where u is not specified at the boundary points $x = x_0, x_1$. The necessary conditions for u to minimize $I(u)$ are that u satisfies the Euler–Lagrange equation

$$\frac{\partial F}{\partial u} - \frac{d}{dx}\frac{\partial F}{\partial u'} = 0,$$

together with the boundary conditions

$$\frac{\partial F}{\partial u'} = 0 \quad (x = x_0, x_1).$$

The latter are known as *natural* boundary conditions because they follow directly from the minimization of the basic integral. If the boundary conditions of the problem consist of neither u specified at the boundary points nor the natural boundary conditions, then the functional to be minimized must be modified in an appropriate manner.

Consider the following modified form of (2.4);

$$I(u) = \int_{x_0}^{x_1} F(x,u,u')dx + [g_1(x,u)]_{x=x_1} - [g_0(x,u)]_{x=x_0},$$

where $g_0(x,u)$ and $g_1(x,u)$ are unspecified functions. The necessary conditions for this modified integral to have a minimum are (Schechter, 1967, p. 28)

$$\frac{\partial F}{\partial u} - \frac{d}{dx}\frac{\partial F}{\partial u'} = 0,$$

together with the boundary conditions

$$\frac{\partial F}{\partial u'} + \frac{\partial g_0}{\partial u} = 0 \quad (x = x_0)$$

and

$$\frac{\partial F}{\partial u'} + \frac{\partial g_1}{\partial u} = 0 \quad (x = x_1).$$

Thus the functions $g_0(x,u)$ and $g_1(x,u)$ can be obtained to suit the boundary conditions of the problem. For example, the variational principle equivalent to the differential problem consisting of the equation

$$u'' + f(x) = 0$$

together with the boundary conditions

$$-u' + \alpha u = 0 \quad (x = x_0)$$

and

$$u' + \beta u = 0 \quad (x = x_1)$$

is based on the functional

$$I(u) = \int_{x_0}^{x_1} [\tfrac{1}{2}u'^2 - f(x)u]\,dx + [\tfrac{1}{2}\beta u^2]_{x=x_1} - [\tfrac{1}{2}\alpha u^2]_{x=x_0}.$$

If we now consider a variational problem in two space dimensions consisting of the minimization of the integral

$$I(u) = \iint_R F(x,y,u,u_x,u_y)\,dxdy,$$

where u is not specified on the boundary of the region R, the necessary conditions for u to minimize $I(u)$ are that u satisfies the differential equation

$$\frac{\partial F}{\partial u} - \frac{\partial}{\partial x}\frac{\partial F}{\partial u_x} - \frac{\partial}{\partial y}\frac{\partial F}{\partial u_y} = 0$$

together with the natural boundary condition

$$\frac{\partial F}{\partial u_x}\frac{dy}{d\sigma} - \frac{\partial F}{\partial u_y}\frac{dx}{d\sigma} = 0$$

on the curve ∂R which encloses the region R. If the normal to ∂R makes an angle α with the x-axis (see Figure 7), then $\cos\alpha = dy/d\sigma$, and $\sin\alpha = -dx/d\sigma$, where σ denotes arc length along the boundary. If this is extended to the case of two dependent functions u and v (Hildebrand,

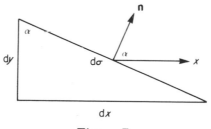

Figure 7

1965, p. 135), the additional Euler–Lagrange equation in v is

$$\frac{\partial F}{\partial v} - \frac{\partial}{\partial x}\frac{\partial F}{\partial v_x} - \frac{\partial}{\partial y}\frac{\partial F}{\partial v_y} = 0,$$

and the additional natural boundary condition is

$$\frac{\partial F}{\partial v_x}\frac{dy}{d\sigma} - \frac{\partial F}{\partial v_y}\frac{dx}{d\sigma} = 0.$$

If the boundary conditions are other than u (and v) specified on the boundary or natural boundary conditions, then the functional again requires modification. This is illustrated in the case of a problem in two space variables with second derivatives present in the integral. Consider the problem of finding the function $u(x, y)$ which gives a stationary value to the functional

$$I(u) = \iint_R F(x, y, u, u_x, u_y, u_{xx}, u_{xy}, u_{yy})dxdy$$

$$+ \int_{\partial R} G(x, y, u, u_\sigma, u_{\sigma\sigma}, u_n)d\sigma$$

where $\partial/\partial\sigma$ and $\partial/\partial n$ are partial differential operators in the directions of the tangent and normal to the curve ∂R. The integration round ∂R is carried out in the sense of an observer walking round ∂R with the region R on his left. The necessary conditions for $I(u)$ to have a minimum value are the Euler–Lagrange equation

$$F_u - \frac{\partial}{\partial x}F_{u_x} - \frac{\partial}{\partial y}F_{u_y} + \frac{\partial^2}{\partial x^2}F_{u_{xx}} + \frac{\partial^2}{\partial x\partial y}F_{u_{xy}} + \frac{\partial^2}{\partial y^2}F_{u_{yy}} = 0$$

together with the boundary conditions

$$\left[F_{u_x} - \frac{\partial}{\partial x}F_{u_{xx}}\right]\frac{dy}{d\sigma} - \left[F_{u_y} - \frac{\partial}{\partial y}F_{u_{yy}}\right]\frac{dx}{d\sigma}$$

$$- \left\{\frac{\partial}{\partial\sigma}(F_{u_{xx}} - F_{u_{yy}})\right\}\frac{dx}{d\sigma}\frac{dy}{d\sigma}$$

$$+ \frac{1}{2}\left\{\frac{\partial}{\partial\sigma}F_{u_{xy}}\left\{\left(\frac{dx}{d\sigma}\right)^2 - \left(\frac{dy}{d\sigma}\right)^2\right\}\right\}$$

$$+ \frac{1}{2}\left\{\left(\frac{\partial}{\partial x}F_{u_{xy}}\right)\frac{dx}{d\sigma} - \left(\frac{\partial}{\partial y}F_{u_{xy}}\right)\frac{dy}{d\sigma}\right\}$$

$$+ G_u - \frac{\partial}{\partial\sigma}G_{u_\sigma} + \frac{\partial^2}{\partial\sigma^2}G_{u_{\sigma\sigma}} = 0 \tag{2.5}$$

and

$$\frac{\partial G}{\partial u_n} + \frac{\partial F}{\partial u_{xx}}\left(\frac{dy}{d\sigma}\right)^2 + \frac{\partial F}{\partial u_{yy}}\left(\frac{dx}{d\sigma}\right)^2 + \frac{\partial F}{\partial u_{xy}}\frac{dx}{d\sigma}\frac{dy}{d\sigma} = 0. \qquad (2.6)$$

The function G is chosen so that the boundary conditions given by (2.5) and (2.6) correspond to the natural boundary conditions of the problem.

2.4 MIXED VARIATIONAL PRINCIPLES

We now return to the small displacement theory of elasticity and to the principle of minimum potential energy which is given in example (2) of Section 2.6. This principle assumes that stress—strain relations and strain—displacement relations are valid and that the kinematic boundary conditions are satisfied.

If the last two conditions are relaxed and are considered as a set of constraints, a modified functional can be written as

$$I(u,v,w) = I_p - \iiint_R \left(\epsilon_x - \frac{\partial u}{\partial x}\right)\alpha_1\,dx$$

$$\cdots$$

$$- \iiint_R \left(\gamma_{zx} - \frac{\partial u}{\partial z} - \frac{\partial w}{\partial x}\right)\alpha_6\,dx$$

$$- \int_{S_d}(u-\bar{u})\beta_1\,d\sigma - \int_{S_d}(v-\bar{v})\beta_2\,d\sigma$$

$$- \int_{S_d}(w-\bar{w})\beta_3\,d\sigma, \qquad (2.7)$$

where I_p is the potential energy functional and $\alpha_1, \ldots, \alpha_6, \beta_1, \ldots, \beta_3$ are Lagrange multipliers. The quantities which can be varied independently are the three displacements, the six strains and the nine Lagrange multipliers. Many mixed variational principles can be obtained from (2.7). These include the Hu—Washizu principle, where

$$\alpha_1 = \sigma_x, \ldots, \alpha_6 = \tau_{zx}, \quad \beta_1 = \sigma_x, \quad \beta_2 = \sigma_y, \quad \beta_3 = \sigma_z,$$

and the Reissner—Hellinger principle, where

$$\alpha_1 = \sigma_x, \ldots, \alpha_6 = \tau_{zx}.$$

The functional associated with the Reissner—Hellinger principle can be written as

$$I_{RH} = I_C - \iiint_R \left(\frac{\partial \sigma_x}{\partial x} + \frac{\partial \tau_{xy}}{\partial y} + \frac{\partial \tau_{zx}}{\partial z} + X \right) u \, dx$$

$$- \iiint_R \left(\frac{\partial \tau_{xy}}{\partial x} + \frac{\partial \sigma_y}{\partial y} + \frac{\partial \tau_{yz}}{\partial z} + Y \right) v \, dx$$

$$- \iiint_R \left(\frac{\partial \tau_{zx}}{\partial x} + \frac{\partial \tau_{yz}}{\partial y} + \frac{\partial \sigma_z}{\partial z} + Z \right) w \, dx$$

$$+ \int_{S_\sigma} (l\sigma_x + m\tau_{xy} + n\tau_{zx} - \bar{X}) u \, d\sigma$$

$$+ \int_{S_\sigma} (l\tau_{xy} + m\sigma_y + n\tau_{yz} - \bar{Y}) v \, d\sigma$$

$$+ \int_{S_\sigma} (l\tau_{zx} + m\tau_{yz} + n\sigma_z - \bar{Z}) w \, d\sigma, \tag{2.8}$$

where I_C is the complementary energy functional.

The Hu—Washizu and Reissner—Hellinger principles are mixed principles and produce stationary but not extremum values. Nevertheless in an approximate method (e.g. finite element method) based on a mixed variational principle, quantities such as displacements and stresses are found with comparable accuracy. This contrasts with a minimum-energy principle where good accuracy can be obtained either in the displacements or the stresses but not in both. For a fuller discussion of mixed variational principles and a complete definition of the notation used in (2.7) and (2.8), the interested reader is referred to Washizu (1968) and Tabarrok (1973).

2.5 TIME-DEPENDENT VARIATIONAL PRINCIPLES

The most basic and important time-dependent variational principle is *Hamilton's* principle from which can be deduced the fundamental equations of a large number of physical phenomena. Hamilton's principle states that the motion of a system from time t_0 to time t_1 is such that the time integral of the difference between the kinetic and potential energies is stationary for the true path. This can be expressed in mathematical terms by defining the integral in terms of the Lagrangian L as

$$I = \int_{t_0}^{t_1} L \, dt = \int_{t_0}^{t_1} (T - V) \, dt,$$

where T, V are the respective kinetic and potential energies of the system, and stating that I is made stationary by the actual motion compared with neighbouring virtual motions. For a system with n generalized coordinates, $q_1, q_2, \ldots, q_n$, the associated Euler–Lagrange equations are

$$\frac{d}{dt}\left(\frac{\partial T}{\partial \dot{q}_r}\right) - \frac{\partial}{\partial q_r}(T - V) = 0 \quad (r = 1, 2, \ldots, n).$$

These are usually referred to as Lagrange's equations of motion for the system.

As a simple example of the foregoing theory applied to a continuum, we consider the case of a flexible string under constant tension τ. The string which is fixed at the ends executes small vibrations about the position of stable equilibrium, which is the interval $0 \leqslant x \leqslant 1$ of the x-axis. If $u(x, t)$ is the displacement perpendicular to the x-axis of a point on the string, then

$$T = \tfrac{1}{2}\rho \int_0^1 \left(\frac{\partial u}{\partial t}\right)^2 dx \quad \text{and} \quad V = \tfrac{1}{2}\rho \int_0^1 c^2 \left(\frac{\partial u}{\partial x}\right)^2 dx,$$

where ρ is the density of the string and $c^2 = \tau/\rho$. The Euler–Lagrange equation is

$$\frac{\partial^2 u}{\partial t^2} = c^2 \frac{\partial^2 u}{\partial x^2},$$

which is the wave equation of the string. Thus the wave equation for the string is equivalent to the requirement that the difference between the total kinetic and potential energies of the string be as small as possible, on average, subject to the initial and boundary conditions of the problem. Other examples of the theory of this section are the vibrating rod, membrane and plate (Courant and Hilbert, 1953, pp. 244–251).

Time-dependent variational principles are not used extensively at present for the numerical solution of evolutionary problems. These are usually solved by semi-discrete Galerkin methods (see Chapter 6).

We now turn our attention to time-dependent *dissipative* systems, and show how variational principles can be constructed for such systems. The method adopted is to introduce an adjoint system with negative friction. The energy lost by the dissipative system is gained by the adjoint system and so the total energy of the two systems is conserved. For an alternative approach using restricted variational principles, the reader is referred to Rosen (1954). As an example consider the one-dimensional oscillator with friction. Its equation of motion is

$$\ddot{x} + k\dot{x} + n^2 x = 0 \quad (k > 0).$$

It is required to find a variational principle which has this equation as its Euler–Lagrange equation. This is impossible, but if we introduce the adjoint oscillator (represented by the coordinate x^*) with negative friction, its equation of motion is

$$\ddot{x}^* - k\dot{x}^* + n^2 x^* = 0.$$

The purely formal Lagrangian

$$L = \dot{x}\,\dot{x}^* - \tfrac{1}{2}k(x^*\,\dot{x} - x\,\dot{x}^*) - n^2 x\,x^*$$

will be seen to give the above two equations of motion as its Euler–Lagrange equations.

Another important example of a dissipative system is the heat diffusion problem. The governing equation for such a problem in one dimension is

$$\frac{\partial u}{\partial t} = \frac{\partial^2 u}{\partial x^2},$$

and we introduce the adjoint problem which is governed by the equation

$$-\frac{\partial u^*}{\partial t} = \frac{\partial^2 u^*}{\partial x^2}.$$

The formal Lagrangian in this case is

$$L = -\frac{\partial u}{\partial x}\frac{\partial u^*}{\partial x} - \frac{1}{2}\left(u^*\frac{\partial u}{\partial t} - u\frac{\partial u^*}{\partial t}\right),$$

which gives the above two equations as its Euler–Lagrange equations.

2.6 DUAL VARIATIONAL PRINCIPLES

So far our variational principles have been one-sided, i.e. the approximate solution always lies above or below the theoretical solution of the variational principle. It is often possible, however, to construct two variational principles for a problem, where the same quantity d is a minimum and maximum with respect to the two principles. If d^u and d^l are approximate solutions of the minimum and maximum principles respectively, then

$$d^l \leqslant d \leqslant d^u,$$

and so we have a practical method of bounding d. It is to be hoped that the quantity d is of physical significance.

Some example will now be given of problems for which dual variational principles can be constructed:

(1) *The classical Dirichlet problem.* Here the functional

$$I(u) = \iint_R \tfrac{1}{2}\{u_x^2 + u_y^2\}\,dxdy$$

is minimized with respect to continuous functions $u(x,y)$ which have piecewise continuous derivatives in the region R and take prescribed values $u = f(\sigma)$ on ∂R, where σ is the arc length of ∂R, the boundary of R. The complementary or dual problem is the functional

$$J(v) = - \iint_R \tfrac{1}{2}\{v_x^2 + v_y^2\}\,dxdy - \int_{\partial R} vf'(\sigma)d\sigma$$

maximized with respect to continuous functions $v(x,y)$ which have piecewise continuous derivatives in R and satisfy natural boundary conditions on ∂R. In this example

$$\min_u I(u) = \max_v J(v) = d.$$

A numerical implementation of this dual principle is given in Section 7.4(c).

Exercise 4 An incompressible inviscid flow is parallel to the x-axis. Calculate $I(u)$ and $J(v)$ for this problem where R is the square region $0 \leqslant x,y \leqslant 1$, and show that the extreme values coincide. (Hint. The functions u and v are the stream function and potential respectively for the flow.)

Exercise 5 Show that the necessary conditions for $J(v)$ to have a maximum consist of the Euler–Lagrange equation

$$v_{xx} + v_{yy} = 0,$$

together with the natural boundary condition

$$v_y \frac{dx}{d\sigma} - v_x \frac{dy}{d\sigma} = f'(\sigma).$$

(2) *Small displacement theory of elasticity* (Washizu, 1968). Consider an isotropic body in three-dimensional space occupying a region R enclosed by surface ∂R. The components of the body forces per unit volume are (X,Y,Z). The surface of the body is divided into two parts, S_σ where the boundary conditions consist of external forces $(\overline{X},\overline{Y},\overline{Z})$ per unit area, and S_d over which the displacements $(\bar{u},\bar{v},\bar{w})$ are given. We have $\partial R = S_\sigma + S_d$. Now the total potential energy is given by

$$I_p = \iiint_R W(\epsilon_x,\epsilon_y,\epsilon_z,\gamma_{yz},\gamma_{zx},\gamma_{xy})dx - \iiint_R (Xu + Yv + Zw)dx$$

$$- \int_{S_\sigma} (\overline{X}u + \overline{Y}v + \overline{Z}w)d\sigma,$$

where

$$W = \frac{E\nu}{2(1 + \nu)(1 - 2\nu)}(u_x + v_y + w_z)^2 + \frac{E}{2(1 + \nu)}(u_x^2 + v_y^2 + w_z^2)$$

$$+ \frac{E}{4(1 + \nu)}[(v_z + w_y)^2 + (w_x + u_z)^2 + (u_y + v_x)^2],$$

with E and ν, Young's modulus and Poisson's ratio, for the material. If the body forces and the surface forces are kept unchanged during variation, I_p is a minimum due to the actual displacements. This is the principle of *minimum potential energy*.

The complementary energy is given by

$$I_c = \iiint_R \varphi(\sigma_x, \sigma_y, \sigma_z, \tau_{yz}, \tau_{zx}, \tau_{xy})\mathrm{dx} - \int_{S_d}(X\bar{u} + Y\bar{v} + Z\bar{w})\mathrm{d}\sigma,$$

where

$$\varphi = \frac{1}{2E}[(\sigma_x + \sigma_y + \sigma_z)^2$$

$$+ 2(1 + \nu)(\tau_{yz}^2 + \tau_{zx}^2 + \tau_{xy}^2 - \sigma_y\sigma_z - \sigma_z\sigma_x - \sigma_x\sigma_y)].$$

If the surface displacements are kept unchanged during variation, I_c is a minimum due to the actual stresses. This is the principle of *minimum complementary energy*.

The quantity which can be bounded conveniently by these two principles is the direct influence coefficient or generalized displacement (Pian, 1970).

Exercise 6   Show that $W = \varphi$ if the following linear stress–strain relations hold:

$$E\epsilon_x = \sigma_x - \nu(\sigma_y + \sigma_z), \quad \tau_{yz} = \frac{E}{2(1 + \nu)}\gamma_{yz},$$

$$E\epsilon_y = \sigma_y - \nu(\sigma_x + \sigma_z), \quad \tau_{zx} = \frac{E}{2(1 + \nu)}\gamma_{zx},$$

$$E\epsilon_z = \sigma_z - \nu(\sigma_x + \sigma_y), \quad \tau_{xy} = \frac{E}{2(1 + \nu)}\gamma_{xy}.$$

Exercise 7   Show that the necessary conditions for the potential energy

$$I_p = \iint_R \left[\frac{E\nu}{2(1 + \nu)(1 - 2\nu)}(u_x + v_y)^2 + \frac{E}{2(1 + \nu)}(u_x^2 + v_y^2)\right.$$

$$\left. + \frac{E}{4(1 + \nu)}(u_y + v_x)^2\right]\mathrm{d}x\mathrm{d}y$$

to have a minimum are the Euler—Lagrange equations

$$(2 - 2\nu)u_{xx} + (1 - 2\nu)u_{yy} + v_{xy} = 0$$

and

$$(2 - 2\nu)v_{yy} + (1 - 2\nu)v_{xx} + u_{xy} = 0$$

in the region R together with the boundary conditions

$$(2 - 2\nu)u_x \cos \alpha + (1 - 2\nu)u_y \sin \alpha + (1 - 2\nu)v_x \sin \alpha$$
$$+ 2\nu v_y \cos \alpha = 0$$

and

$$2\nu u_x \sin \alpha + (1 - 2\nu)u_y \cos \alpha + (1 - 2\nu)v_x \cos \alpha$$
$$+ (2 - 2\nu)v_y \sin \alpha = 0$$

on ∂R (see Figure 7.)

(3) *Compressible flow* (Sewell, 1969).   The respective volume integrands which appear in the dual variational principles are the pressure p and the quantity $p + \rho v^2$, where ρ is the density and v is the velocity of the fluid. Here

$$p = p(v_i, h, \eta),$$

where v_i ($i = 1,2,3$) are the velocity components, h and η are the total energy and entropy per unit mass respectively. The results

$$\frac{\partial p}{\partial v_i} = -Q_i \quad (i = 1,2,3), \qquad \frac{\partial p}{\partial h} = \rho, \qquad \frac{\partial p}{\partial \eta} = -\rho T$$

follow, where $Q_i = \rho v_i$ ($i = 1,2,3$) with ρ the density and T the temperature. The function

$$P = P(Q_i, h, \eta) = \sum_{i=1}^{3} Q_i v_i + p$$

is introduced, where

$$\frac{\partial P}{\partial Q_i} = v_i \quad (i = 1,2,3), \qquad \frac{\partial P}{\partial h} = \rho, \qquad \frac{\partial P}{\partial \eta} = -\rho T.$$

The dual variational principles involving p and P respectively can be strengthened to extremum principles for particular types of compressible flow.

Two accounts attempting to unify dual principles are due to Sewell (1969) and Arthurs (1970). The former uses Legendre or involutory transformations, and the latter uses the *canonical theory* of the Euler—Lagrange equation. An excellent account of dual variational principles in general can be found in Noble and Sewell (1972).

Chapter 3

Methods of Approximation

The variational formulation, with the weaker continuity requirements, lends itself naturally to approximate methods of solution usually referred to as *direct methods* (Courant and Hilbert, 1953, p. 174; Nečas, 1967). Such methods transform the problem into one involving the stationary points of a function of a finite number of real variables.

In the present chapter, however, we are only interested in the family of direct methods known generally as finite element methods. A description of the various alternative forms of finite element methods are given, together with a few examples of the wide range of problems to which they can be applied; additional examples are given in Chapter 7. The accuracy and convergence of the methods are discussed in Chapter 5.

3.1 RITZ METHODS

The classical direct method is usually attributed to the Swiss mathematician W. Ritz (1878–1909). If we require the solution of the variational problem

$$\delta I(v) = 0 \qquad (v(\mathbf{x}) \in \mathcal{H}), \tag{3.1}$$

where $\mathbf{x} = (x_1, \ldots, x_m)$ and $\mathcal{H}$ is the space of admissible functions, an approximate solution can be obtained by restricting the function to some N-dimensional subspace $K_N \subset \mathcal{H}$. It follows immediately that if the stationary point in (3.1) is an extremum, we have a bound on that extremum value. That is, when

$$\delta I(v_0) = 0$$

if and only if

$$I(v_0) = \min_{v \in \mathcal{H}} I(v) = d,$$

then

$$I(V) \geqslant d$$

for all $V \in K_N$.

If we assume that the functions $\varphi_i(\mathbf{x})$ $(i = 1, \ldots, N)$ form a basis for the subspace K_N, then the approximate solution

$$U(\mathbf{x}) = \sum_{i=1}^{N} \alpha_i \varphi_i(\mathbf{x}),$$

is such that $I(U)$ is stationary with respect to each of the parameters α_i, $(i = 1, 2, \ldots, N)$ and so we obtain the system of equations

$$\frac{\partial}{\partial \alpha_i} I \left(\sum_{j=1}^{N} \alpha_j \varphi_j \right) = 0 \quad (i = 1, \ldots, N).$$

As an example of this approach, let us apply the Ritz method to a particular problem: Let the equation

$$\frac{\partial^2 u}{\partial x^2} + \frac{\partial^2 u}{\partial y^2} - 2 = 0 \tag{3.2}$$

be valid for $-\tfrac{1}{2}\pi < x, y < \tfrac{1}{2}\pi$ subject to

$$u(x, \pm\tfrac{1}{2}\pi) = 0 \qquad (|x| \leqslant \tfrac{1}{2}\pi)$$

and $\tag{3.2a}$

$$u(\pm\tfrac{1}{2}\pi, y) = 0 \qquad (|y| \leqslant \tfrac{1}{2}\pi).$$

A finite element approximation is required with piecewise bilinear functions, using the Ritz method applied to the corresponding variational principle

$$\delta I(v) = 0,$$

with

$$I(v) = \iint_{R} [\tfrac{1}{2}(v_x^2 + v_y^2) + 2v]\,dxdy, \tag{3.3}$$

where

$$R = (-\tfrac{1}{2}\pi, \tfrac{1}{2}\pi) \times (-\tfrac{1}{2}\pi, \tfrac{1}{2}\pi).$$

In the previous chapter we show that if the value of the solution is prescribed on the boundary (*Dirichlet boundary condition*) then this condition must be imposed on the space $\mathcal{H}$, whereas this is not necessary if the natural boundary conditions are given. It follows that in this problem, it is necessary to restrict the space $\mathcal{H}$ to functions that satisfy the boundary condition $v = 0$.

The approximating functions are defined on the region R, partitioned into $(T + 1)^2$ square elements, by means of T equally spaced internal grid lines parallel to each axis (see Figure 3). Then the $N (=T^2)$ basis functions $\varphi_{ij}(x,y)$ $(i,j = 1, \ldots, T)$, for subspace K_N, are defined in Chapter 1 (Exercise 6) and clearly belong to $\mathcal{H}$ as they all vanish on the boundary.

The approximate solution is then of the form

$$U(x,y) = \sum_{i,j=1}^{T} U_{ij}\varphi_{ij}(x,y), \tag{3.4}$$

where U_{ij} is the approximate solution at the point (x_i,y_j). The stationary is then given by

$$\frac{\partial}{\partial U_{ij}} I\left(\sum_{k,l=1}^{T} U_{kl}\varphi_{kl}(x,y)\right) = 0 \quad (i,j = 1, \ldots, T). \tag{3.5}$$

The functional I is given by (3.3) and thus (3.5) leads to

$$\sum_{k,l=1}^{T} U_{kl} \iint_R \left\{\left(\frac{\partial\varphi_{kl}}{\partial x}\right)\left(\frac{\partial\varphi_{ij}}{\partial x}\right) + \left(\frac{\partial\varphi_{kl}}{\partial y}\right)\left(\frac{\partial\varphi_{ij}}{\partial y}\right)\right\} dxdy$$

$$+ 2 \iint_R \varphi_{ij}\, dxdy = 0.$$

Straightforward evaluation of the integrals leads to the equation

$$3U_{ij} - \frac{1}{3} \sum_{k=i-1}^{i+1} \sum_{l=j-1}^{j+1} U_{kl} + 2h^2 = 0 \quad (i,j = 1, \ldots, T),$$

where $U_{kl} = 0$ if $k,l = 0, T+1$. This equation can be written in terms of difference operators as

$$\{\delta_x^2 I_y + \delta_y^2 I_x\} U_{ij} - 2h^2 = 0 \quad (i,j = 1, \ldots, T),$$

where δ_x^2, δ_y^2 are second-central-difference operators and I_x, I_y are 'Simpson's Rule' operators defined by

$$I_x U_{ij} = \tfrac{1}{6}[U_{i-1\,j} + 4U_{ij} + U_{i+1\,j}]$$

and similarly for I_y. The approximate solution at the grid points shown in Figure 8, is given in Table 1; owing to symmetry only one-eighth of the region needs to be displayed. The theoretical solution is

$$u(x,y) = -(\tfrac{1}{2}\pi)^2 + x^2 + \frac{8}{\pi} \sum_{k=1}^{\infty} \frac{(-1)^{k+1}}{(2k-1)^3} \frac{\cosh(2k-1)y}{\cosh(2k-1)\tfrac{1}{2}\pi} \cos(2k-1)x.$$

If equation (3.2) is valid in the region $-\tfrac{1}{2}\pi < x, y < \tfrac{1}{2}\pi$ subject to the natural boundary conditions

$$\frac{\partial u(x, \pm\tfrac{1}{2}\pi)}{\partial y} = 0 \quad (|x| \leqslant \tfrac{1}{2}\pi)$$

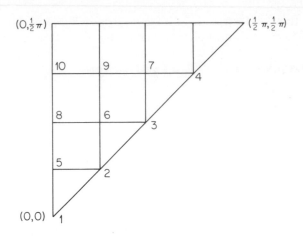

Figure 8

and

$$\frac{\partial u(\pm\frac{1}{2}\pi, y)}{\partial x} = 0 \quad (|y| \leqslant \frac{1}{2}\pi),$$

the space of admissible functions $\mathcal{H}$, also contains functions that take non-zero values on the boundary. The approximating subspace should also contain such functions, so we add basis functions $\varphi_{ij}(x,y)$ corresponding to the boundary points (x_i, y_j) where $x_i, y_j = \pm\frac{1}{2}\pi$. Such functions are non-zero in at most two elements, as shown in Figure 9(a) rather than four elements, as shown in Figure 9(b).

Table 1

		Solution		
Point	$T = 3$	$T = 7$	$T = 15$	Exact
1	−1.534	−1.473	−1.459	−1.454
2		−1.321	−1.308	−1.304
3	−0.950	−0.907	−0.897	−0.894
4		−0.370	−0.362	−0.359
5		−1.394	−1.381	−0.376
6		−1.089	−1.078	−1.075
7		−0.566	−0.559	−0.556
8	−1.278	−1.146	−1.135	−1.132
9		−0.666	−0.660	−0.658
10		−0.698	−0.692	−0.690

Corner   Side boundary

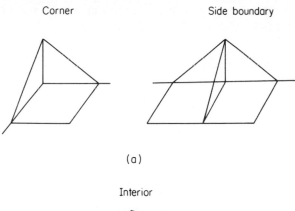

(a)

Interior

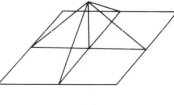

(b)

Figure 9

Exercise 1   Using the Ritz method with piecewise bilinear basis
functions, calculate the coefficients in the equation

$$\frac{\partial}{\partial U_{ij}} I = 0,$$

assuming that the point (x_i, y_j) is (a) a corner point, and (b) a side
point, where I is given by (3.3), together with the *natural* boundary
conditions.

Systems of equations

The Ritz method can be applied to a system of partial differential
equations, as for example in the small displacement theory of elasticity.
In general if we take approximations of the form

$$U = \sum_{i=1}^{N_1} \alpha_i \varphi_i$$

and   (3.6)

$$V = \sum_{l=1}^{N_2} \beta_l \psi_l$$

and obtain the stationary point of a functional $I(U,V)$ with respect to U and V, this leads to the set of $N_1 + N_2$ equations

$$\frac{\partial}{\partial \alpha_i} I\left(\sum_{j=1}^{N_1} \alpha_j \varphi_j, \sum_{k=1}^{N_2} \beta_k \psi_k\right) = 0 \quad (i = 1, \ldots, N_1) \tag{3.7}$$

and

$$\frac{\partial}{\partial \beta_l} I\left(\sum_{j=1}^{N_1} \alpha_j \varphi_j, \sum_{k=1}^{N_2} \beta_k \psi_k\right) = 0 \quad (l = 1, \ldots, N_2).$$

3.2 BOUNDARY CONDITIONS

We have shown how two different types of problems can be solved by finite element method. First we considered the boundary condition $u = 0$ or the *homogeneous Dirichlet* boundary condition, when it was necessary for all admissible functions to satisfy this condition. Then we considered the condition $\partial u/\partial n = 0$, where n is the outward normal direction, this is the *homogeneous Neumann* boundary condition, when no restrictions were imposed on the admissible functions as the boundary conditions were the natural boundary conditions for the particular functional (3.3).

Dirichlet boundary conditions

In general it is necessary to introduce Dirichlet boundary conditions as additional conditions on the approximating functions. The easiest case to deal with, is where we have to solve the linear differential equation

$$Au = f \tag{3.8}$$

in some region $R \subset \mathbb{R}^m$, subject to the condition

$$u = g$$

on the boundary ∂R, where g is a smooth function with a straight-forward analytic continuation into R. That is, there exists a smooth function w such that

$$w = g$$

on ∂R, and Aw is defined throughout R. In such cases, it is possible to consider approximate solutions of the form

$$U = w + \sum_{i=1}^{N} \alpha_i \varphi_i, \tag{3.9}$$

where all the φ_i are zero on ∂R (Synge, 1957). As functions of the form

(3.9) define a *linear manifold* but not a linear space, it is often more convenient, for the mathematical analysis, to transform the problem slightly. After the transformation the approximation problem has the structure of the classical approximations described in Sections 1.3 and 3.5. It should be emphasized that the numerical calculation is not affected by this transformation; only the mathematical formulation has been altered. The foregoing is illustrated in the following exercise.

Exercise 2   Let equation (3.8) be valid for $-\frac{1}{2}\pi < x, y < \frac{1}{2}\pi$ with $f = 0$, $g = -\frac{1}{2}(x^2 + y^2)$ and

$$A = -\frac{\partial^2}{\partial x^2} - \frac{\partial^2}{\partial y^2}.$$

Let

$$w(x,y) = g(x,y) = -\frac{1}{2}(x^2 + y^2) \tag{3.10}$$

and then solve

$$A[v(x,y) - \frac{1}{2}(x^2 + y^2)] = 0,$$

that is,

$$\frac{\partial^2 v}{\partial x^2} + \frac{\partial^2 v}{\partial y^2} - 2 = 0 \tag{3.11}$$

subject to $v = 0$ on the boundary. Using the solution of (3.11) given in Table 1, or otherwise, calculate the solution at the grid points in Figure 8 and compare this approximation with the theoretical solution

$$u(x,y) = -(\tfrac{1}{2}\pi)^2 - \frac{1}{2}(y^2 - x^2)$$

$$+ \frac{8}{\pi} \sum_{k=1}^{\infty} \frac{(-1)^{k+1}}{(2k-1)^3} \frac{\cosh(2k-1)y}{\cosh(2k-1)\frac{1}{2}\pi} \cos(2k-1)x.$$

Alternative boundary value approximation

There are unfortunately many practical problems in which the Dirichlet boundary conditions are not sufficiently smooth to ensure that a simple analytic continuation is possible. In such problems it may be convenient to approximate the boundary data in some way using basis functions φ_i that are non-vanishing on the boundary. In this way we construct an approximate solution

$$U = \sum_{i=1}^{N} \alpha_i \varphi_i,$$

where some of the α_i are fixed by the boundary data. They could be chosen, for example, so that the approximate solution interpolates the boundary conditions. The remainder, corresponding to the internal nodes, are computed by the Ritz approximation procedure. This type of approximation does not fit into a classical formulation of the finite element method through the calculus of variations, but it is shown later, in Chapter 5, that when performed correctly this does not reduce the accuracy of the approximation.

Another approach that has come into prominence (Bramble and Schatz, 1970; Babuska, 1973; Nitsche, 1971) is to introduce inhomogeneous Dirichlet boundary conditions as a so-called penalty functional. By incorporating the boundary conditions into the functional it is possible to remove all restrictions on the approximating subspace K_N. If, for example, the differential equation (3.2) is given, subject to the condition

$$u = g$$

on the boundary, we could use the functional

$$J_\lambda(v) = \iint_R \left\{ \frac{1}{2} \left[\left(\frac{\partial v}{\partial x} \right)^2 + \left(\frac{\partial v}{\partial y} \right)^2 \right] + 2v \right\} dx\, dy$$

$$+ \frac{1}{2}\lambda \int_{\partial R} (v - g)^2 \, d\sigma. \tag{3.12}$$

Clearly if there is no restriction on the choice of the approximating function v, the functional (3.12) corresponds to (3.2) subject to the natural boundary condition

$$u + (\lambda^{-1}) \frac{\partial u}{\partial n} = g$$

on ∂R. Inhomogeneous Dirichlet boundary conditions are mentioned again later in this chapter, when the method of *least squares* is introduced.

Inhomogeneous Neumann boundary conditions or mixed boundary conditions can be incorporated into a functional analogous to (3.12) without placing any restriction on the approximating subspace.

3.3 THE KANTOROVICH METHOD (OR SEMI-DISCRETE METHOD)

An alternative use of direct methods of approximation is to define the approximate solution

$$U = \sum_{i=1}^{N} \alpha_i \varphi_i,$$

where the unknown coefficients α_i ($i = 1, \ldots, N$) are no longer scalars, but are instead unknown functions of *one* of the independent variables.

Without loss of generality we can take this variable to be x_1. The functions φ_i ($i = 1, \ldots, N$) are then taken as functions of the remaining $m - 1$ variables $x_2, \ldots, x_m$, hence

$$U(x_1, x_2, \ldots, x_m) = \sum_{i=1}^{N} \alpha_i(x_1)\varphi_i(x_2, \ldots, x_m).$$

This method is often associated with the name of L. V. Kantorovich (Kantorovich, 1933); it forms the basis for most of the finite element solutions for time-dependent problems considered in Chapter 6. When applied to boundary value problems, the Kantorovich method is very similar to the well known *method of (straight) lines* (Berezin and Zhidkov, 1965).

As a simple example let us consider how the Kantorovich method would be used to obtain an approximate solution of (3.2) subject to the boundary conditions (3.2a). The approximating subspace K_N is chosen to contain functions of y only, and from (3.2a) such functions will satisfy

$$\varphi_i(-\tfrac{1}{2}\pi) = \varphi_i(\tfrac{1}{2}\pi) = 0 \quad (i = 1, 2, \ldots, N).$$

The approximate solution of the form

$$V(x,y) = \sum_{i=1}^{N} \alpha_i(x)\varphi_i(y), \tag{3.13}$$

is calculated by minimizing the functional $I(v)$, given by (3.3), with respect to the *undetermined functions* α_i ($i = 1, 2, \ldots, N$). To do this it is first necessary to rewrite the functional as

$$I\left(\sum_{i=1}^{N} \alpha_i \varphi_i\right) = J[\alpha_1, \alpha_2, \ldots, \alpha_N]$$

$$= \int_{-\frac{1}{2}\pi}^{\frac{1}{2}\pi} F(\alpha_1, \ldots, \alpha_N)\,dx,$$

where

$$F(\alpha_1, \ldots, \alpha_N) = \sum_{i=1}^{N}\sum_{j=1}^{N} \frac{1}{2}\left\{\alpha_i\alpha_j \int_{-\frac{1}{2}\pi}^{\frac{1}{2}\pi} \frac{d\varphi_i}{dy}\frac{d\varphi_j}{dy}\,dy\right.$$

$$\left. + \frac{d\alpha_i}{dx}\frac{d\alpha_j}{dx}\int_{-\frac{1}{2}\pi}^{\frac{1}{2}\pi} \varphi_i\varphi_j\,dy\right\} + \sum_{i=1}^{N}\alpha_i\int_{-\frac{1}{2}\pi}^{\frac{1}{2}\pi} 2\varphi_i\,dy. \tag{3.14}$$

It is possible to consider the functional $I(v)$ in this manner, since the functions φ_i ($i = 1, 2, \ldots, N$) are known and the integrals in (3.14) can be evaluated explicitly. In order to obtain a stationary value of the functional $J[\alpha_1, \ldots, \alpha_N]$, we follow the procedure outlined in the previous chapter, and derive the *Euler—Lagrange equations*, corresponding to variations of each of the $\alpha_i(x)$ $(i = 1, \ldots, N)$. This is necessary as the coefficients are no longer simple scalars. Thus the α_i $(i = 1, \ldots, N)$ are defined by the N equations

$$\frac{\partial F}{\partial \alpha_i} - \frac{d}{dx}\left(\frac{\partial F}{\partial \alpha_i'}\right) = 0 \quad (i = 1, \ldots, N),$$

where $'$ denotes differentiation with respect to x and where $F(\alpha_1, \ldots, \alpha_N)$ is given by (3.14). Therefore the unknown functions α_i $(i = 1, \ldots, N)$ in the approximate solution (3.13) are given by the system of differential equations

$$\sum_{j=1}^{N} \{\alpha_j c_{ij} - \alpha_j'' d_{ij}\} = b_i \quad (i = 1, \ldots, N), \tag{3.15}$$

where

$$c_{ij} = \int_{-\frac{1}{2}\pi}^{\frac{1}{2}\pi} \frac{d\varphi_i}{dy} \frac{d\varphi_j}{dy} \, dy,$$

$$d_{ij} = \int_{-\frac{1}{2}\pi}^{\frac{1}{2}\pi} \varphi_i \varphi_j \, dy$$

and

$$b_i = -\int_{-\frac{1}{2}\pi}^{\frac{1}{2}\pi} 2\varphi_i \, dy,$$

subject to

$$\alpha_i(-\tfrac{1}{2}\pi) = \alpha_i(\tfrac{1}{2}\pi) = 0 \quad (i = 1, \ldots, N).$$

Exercise 3   Solve the system of equations (3.15) with the functions φ_i $(i = 1, \ldots, N)$ given as piecewise linear functions defined by (1.3) on the interval $[-\frac{1}{2}\pi, \frac{1}{2}\pi]$ with a uniform partition. Compare the results obtained for various values of N with those given in Table 1.

The semi-discrete method can be applied equally well with inhomogeneous Dirichlet boundary conditions, or natural boundary conditions. The procedure becomes complicated when additional boundary integrals have to be incorporated in the functional.

Exercise 4   Construct the functional $J[\alpha_1, \ldots, \alpha_N]$ that would be used to obtain the solution of equation (3.2) in the region $-\frac{1}{2}\pi < x,y < \frac{1}{2}\pi$ subject to the condition $u = g$ on the boundary.

In general, the semi-discrete method is a useful approach for boundary value problems only when the resultant one-dimensional problem can be solved exactly in a straightforward manner. There are a few such examples of the method of Elsgolc (1961) p. 168f. Applications to *initial value problems* are far more significant, and they will be treated at length in Chapter 6.

3.4 GALERKIN METHODS

In this chapter so far, we have considered a number of methods for approximating the solution $u(\mathbf{x})$, of the linear differential equation

$$Au = f, \tag{3.16}$$

for $\mathbf{x} \in \mathbb{R}^m$, subject to certain boundary conditions. It was assumed for all these methods that the differential operator A satisfies conditions such that the function u is the solution of the variational problem

$$\delta I(u) = 0 \tag{3.17}$$

for some functional $I(u)$. In such circumstances it can be shown (Mikhlin, 1964) that the functional can be written in the form

$$I(u) = \tfrac{1}{2}(Au,u) - (u,f), \tag{3.18}$$

where

$$(u,v) = \iint_R u(\mathbf{x})v(\mathbf{x})\mathrm{d}\mathbf{x}.$$

The extension of this result to non-linear operators has been considered in detail by Vainberg (1964). In several of the examples given in the preceding sections of this chapter the operator

$$A = -\frac{\partial^2}{\partial x^2} - \frac{\partial^2}{\partial y^2}$$

has been used, together with the functional

$$I(u) = \iint_R \tfrac{1}{2}(u_x^2 + u_y^2)\mathrm{d}x\mathrm{d}y - \iint_R uf\,\mathrm{d}x\mathrm{d}y. \tag{3.19}$$

This expression can be obtained, subject to certain boundary conditions, from the standard form (3.18), by applying Green's theorem (Courant and Hilbert, 1953, pp. 279f)

Exercise 5 Find the boundary conditions on the function $u(x,y)$, for which the transformation of (3.18) to (3.19) is valid.

The use of integration by parts — that is, Green's theorem — to transform the functional from the standard form (3.18) to one requiring less continuity of the admissible functions $u(\mathbf{x})$, is one of the keys to the success of the finite element method (Prenter, 1975, pp. 229f).

The Ritz approximation

$$U(\mathbf{x}) = \sum_{i=1}^{N} \alpha_i \varphi_i(\mathbf{x})$$

to the solution of the variational problem (3.17), can be written as the solution of

$$\frac{\partial}{\partial \alpha_i} I(U) = (AU, \varphi_i) - (\varphi_i, f) = 0 \quad (i = 1, 2, \ldots, N), \tag{3.20}$$

only when the operator A is of the required form. But even if this is not true, and (3.20) is not valid, the system of equations

$$(AU, \varphi_i) - (\varphi_i, f) = (AU - f, \varphi_i) = 0 \quad (i = 1, 2, \ldots, N) \tag{3.21}$$

still defines an approximate solution $U(\mathbf{x})$ of (3.16). In fact the system (3.21) can be used even if the operator is non-linear. Thus it is possible to use the finite element method on a much wider class of problems than those corresponding to variational principles. It is still however desirable that (3.21) can be transformed by integration by parts to reduce the degree of continuity required of the functions φ_i. Such approximations are usually associated with the name of the Russian mathematician B. G. Galerkin (1871—1945). In many Russian textbooks it is referred to as the Bubnov—Galerkin method (Mikhlin and Smolitsky, 1967; Vulikh, 1963).

Weak solutions

In some mathematical texts, a function $u(\mathbf{x})$ that satisfies the linear differential equation

$$Au = f \quad (\mathbf{x} \in R)$$

is called a *classical solution*. This is to distinguish it from a *weak solution* which satisfies

$$(Au, v) = (f, v) \quad \text{(for all } v \in \mathcal{H}) \tag{3.22a}$$

or

$$(u, A^*v) = (f, v) \quad \text{(for all } v \in \mathcal{H}_1). \tag{3.22b}$$

The space $\mathcal{H}$ contains all measurable admissible functions that vanish on the boundary ∂R; such functions are sometimes said to have *compact support*. The space $\mathcal{H}_1 \subset \mathcal{H}$ contains only those admissible functions for which A^*v is measurable.

It follows that if the operator A is of order $2k$, then the weak form (3.22a) requires the solution u to have measurable derivatives of order $2k$; in the Sobolev space notation of Chapter 5 this can be written as $u \in \mathcal{H}_2^{(2k)}(R)$. The second weak form (3.22b) however, requires only that $u \in \mathcal{L}_2(R)$; the situation is reversed for the *test functions* v. The additional requirements that they should vanish on the boundary is usually — as in Chapter 5 — introduced into the notation as $v \in \overset{\circ}{\mathcal{H}}_2^{(2k)}(R)$.

From a computational point of view, the most useful weak form —

which is referred to hereafter as the *Galerkin form* — is derived from either (3.22a) or (3.22b) by k integrations by parts. This Galerkin form has the minimum continuity requirements in the sense that $u \in \mathcal{H}_2^{(k)}(R)$ and $v \in \mathring{\mathcal{H}}_2^{(k)}(R)$. It is written in the standard form

$$a(u,v) = (f,v) \quad \text{(for all } v \in \mathcal{H}_2), \tag{3.23}$$

where the new space of test functions $\mathcal{H}_2$ clearly satisfies

$$\mathcal{H}_1 \subset \mathcal{H}_2 \subset \mathcal{H}.$$

For example, the Galerkin form for the differential operator

$$-\frac{\partial^2}{\partial x^2} - \frac{\partial^2}{\partial y^2}$$

leads to

$$a(u,v) = \iint_R \left[\left(\frac{\partial u}{\partial x}\right)\left(\frac{\partial v}{\partial x}\right) + \left(\frac{\partial u}{\partial y}\right)\left(\frac{\partial v}{\partial y}\right) \right] \mathrm{d}x\mathrm{d}y,$$

where in this case $k = 1$. It follows from Chapter 2 that the Galerkin form for the differential operator

$$\frac{\partial^4}{\partial x^4} + 2\frac{\partial^4}{\partial x^2 \partial y^2} + \frac{\partial^4}{\partial y^4}$$

leads to

$$a(u,v) = \iint_R \left(\frac{\partial^2 u}{\partial x^2} + \frac{\partial^2 u}{\partial y^2}\right)\left(\frac{\partial^2 v}{\partial x^2} + \frac{\partial^2 v}{\partial y^2}\right) \mathrm{d}x\mathrm{d}y.$$

It can be shown that if sufficient attention is given to the boundary conditions, the weak solution is unique and it is also a classical solution: one such result is given in Chapter 5 as Theorem 5.2. The existence and uniqueness of weak solutions is also studied in Bers, John and Schechter (1964) p. 138, Nečas (1967) p. 23 and Lions and Magenes (1972) Section 2.9. In this book it will be assumed that if $\mathcal{H}_1$, $\mathcal{H}_2$ and $\mathcal{H}$ are chosen correctly then the solutions of (3.22a), (3.22b) and (3.23) are identical and that their solution is the required classical solution. In general the only weak form that is considered is the Galerkin form. The Galerkin approximation U satisfies

$$a(U,\varphi_i) = (f,\varphi_i) \quad (i = 1, \ldots, N). \tag{3.24}$$

Hence as the functions φ_i span the finite-dimensional space K_N, an equivalent statement of this system of equations is

$$a(U,V) = (f,V) \quad \text{(for all } V \in K_N). \tag{3.24a}$$

Note that it is not necessary for u (or U) to be in the energy space, this is only a requirement on v (or V). Thus if the boundary conditions are inhomogeneous Dirichlet conditions, it may be appropriate to use an approximation of the form

$$U(\mathbf{x}) = W(\mathbf{x}) + \sum_{i=1}^{N} \alpha_i \varphi_i(\mathbf{x})$$

where the function W satisfies the boundary conditions.

In the Galerkin equations, it is natural to assume that the approximation U and the test functions V are defined in terms of the same set of basis functions φ_i ($i = 1, \ldots, N$). By so doing, we ensure that in problems for which both Ritz and Galerkin approximations can be defined the two methods lead to alternative formulations of the same solution. A different approximation, in terms of the same basis functions $\varphi_i \in K_N$, is obtained if the test functions are defined in terms of some $\psi_i \in L_N$ ($i = 1, \ldots, N$); where K_N and L_N are different subspaces of $\mathscr{H}$.

The method of least squares can be considered as a method of this type with

$$\psi_i = A\varphi_i \quad (i = 1, \ldots, N).$$

The use of two sets, $\{\varphi_i\}$ and $\{\psi_i\}$, of disimilar basis functions to define an approximation of the form (3.25) by

$$a(U, \psi_i) = (f, \psi_i) \quad (i = 1, \ldots, N), \tag{3.25}$$

has lead various authors to develop so-called *conjugate approximations* (Oden, 1972, and references therein) and the so-called *method of weighted residuals* (Finlayson and Scriven, 1967, and references therein).

It is shown by example in Section 7.4(E), that in certain practical problems there are definite numerical advantages in choosing the test functions to be of a different form to the approximate solution. It is not possible here, or in Chapter 5, to provide a discussion of the theoretical benefits of such methods; for this, the reader should consult the references.

The Galerkin method can be applied to non-linear problems, but it is only in special cases that it is possible to derive a weak form that has reduced continuity requirements. One example for which there is such a reduction is the non-linear equation

$$\frac{\partial}{\partial x}\left(p(u)\,\frac{\partial u}{\partial x} \right) + \frac{\partial}{\partial y}\left(q(u)\,\frac{\partial u}{\partial y} \right) + f(x,y) = 0.$$

This leads to a Galerkin form

$$a(U, \varphi_i) = \iint_R \left\{ p(U)\left(\frac{\partial U}{\partial x} \right)\left(\frac{\partial \varphi_i}{\partial x} \right) + q(U)\left(\frac{\partial U}{\partial y} \right)\left(\frac{\partial \varphi_i}{\partial y} \right) \right\} \mathrm{d}x\mathrm{d}y$$

$$= (f, \varphi_i) \quad (i = 1, \ldots, N).$$

Adjoint formulations

It is *possible*, to extend the variational calculus to the wider class of problems for which Galerkin methods are available. In the previous chapter, a formal Lagrangian is presented for a *dissipative system*, by introducing an adjoint problem. If approximate solutions are constructed for both the *primary problem* and the *adjoint problem*, the Galerkin approximation is obtained from a stationary point of this Lagrangian. In fact the equations (3.24) appear as necessary conditions for a stationary value.

As a simple example, let us consider the equation

$$\frac{\partial^2 u}{\partial x^2} + \frac{\partial^2 u}{\partial y^2} + 2\frac{\partial u}{\partial y} = 0 \tag{3.26}$$

to be valid in some region $R \subset \mathbb{R}^2$, subject to the condition $u = 0$ on the boundary ∂R. We shall call this the primary problem. The adjoint problem then consists of the adjoint equation

$$\frac{\partial^2 u^*}{\partial x^2} + \frac{\partial^2 u^*}{\partial y^2} - 2\frac{\partial u^*}{\partial y} = 0 \tag{3.27}$$

in R, subject to the equivalent condition $u^* = 0$ on the boundary.

It is easy to verify that both (3.26) and (3.27) arise as Euler–Lagrange equations, corresponding to a variation of the functional

$$I(u,u^*) = \iint_R \left\{ \frac{\partial u}{\partial x}\frac{\partial u^*}{\partial x} + \frac{\partial u}{\partial y}\frac{\partial u^*}{\partial y} + u\frac{\partial u^*}{\partial y} - u^*\frac{\partial u}{\partial y} \right\} dxdy. \tag{3.28}$$

It should be remembered however, that the primary equation (3.26) is given by

$$\delta_{u^*} I(u,u^*) = 0 \tag{3.29a}$$

and not by

$$\delta_u I(u,u^*) = 0, \tag{3.29b}$$

as the latter leads to the adjoint equation.

If we seek an approximate solution for the primary problem, of the form

$$U - \sum_{i=1}^{N} \alpha_i \varphi_i,$$

it is necessary to introduce an approximate solution for the adjoint problem, of the form

$$U^* = \sum_{i=1}^{N} \beta_i \varphi_i;$$

another form of adjoint approximation would lead to (3.25) rather than (3.24). These approximations can be found from the stationary value of the functional (3.28) with respect to the unknown coefficients α_i and β_i. Thus we replace (3.29a) by

$$\frac{\partial}{\partial \beta_i} I(U, U^*) = 0 \quad (i = 1, \dots, N) \tag{3.30a}$$

and (3.29b) by

$$\frac{\partial}{\partial \alpha_i} I(U, U^*) = 0 \quad (i = 1, \dots, N). \tag{3.30b}$$

If we consider the functional $I(U, U^*)$ given by (3.28), the approximate solution of the primary equation is given by the set of equations

$$\iint_R \left\{ \left(\frac{\partial U}{\partial x}\right) \left(\frac{\partial \varphi_i}{\partial x}\right) + \left(\frac{\partial U}{\partial y}\right) \left(\frac{\partial \varphi_i}{\partial y}\right) + U\left(\frac{\partial \varphi_i}{\partial y}\right) - \left(\frac{\partial U}{\partial y}\right) \varphi_i \right\} dxdy = 0$$
$$(i = 1, \dots, N). \tag{3.31}$$

The more direct formulation of the Galerkin approximation is derived by using the weak form of (3.26). It is easy to verify that the Galerkin form of (3.26) is

$$a(u, v) = 0 \quad \text{(for all } v \in \mathcal{H}),$$

where

$$a(u, v) = \iint_R \left\{ \left(\frac{\partial u}{\partial x}\right) \left(\frac{\partial v}{\partial x}\right) + \left(\frac{\partial u}{\partial y}\right) \left(\frac{\partial v}{\partial y}\right) + u\left(\frac{\partial v}{\partial y}\right) - \left(\frac{\partial u}{\partial y}\right) v \right\} dxdy,$$

it then follows immediately that the Galerkin approximation given by

$$a(U, \varphi_i) = 0 \quad (i = 1, \dots, N) \tag{3.32}$$

satisfies (3.31). Thus, it has been shown in this particular problem how the same set of equations can be formulated in two mathematically different ways; from a weak solution to the original differential equation, or from a stationary value of some functional. The adjoint problem, which is required in the definition of the appropriate functional, is a mathematical concept with (in general) no physical significance. Therefore for the remainder of this book, Galerkin methods are developed from weak solutions with only a passing reference to the alternative formulation.

As an example let equation (3.26), be valid in the region

$-\tfrac{1}{2}\pi < x,y < \tfrac{1}{2}\pi$, subject to the boundary conditions

$$u(\pm\tfrac{1}{2}\pi, y) = 0 \qquad\qquad (|\,y\,| \leqslant \tfrac{1}{2}\pi),$$

$$u(x, -\tfrac{1}{2}\pi) = 0 \qquad\qquad (|\,x\,| \leqslant \tfrac{1}{2}\pi)$$

and

$$u(x, \tfrac{1}{2}\pi) = (\tfrac{1}{2}\pi)^2 - x^2 \qquad (|\,x\,| \leqslant \tfrac{1}{2}\pi).$$

(3.33)

Then the Galerkin finite element method can be applied to this problem to derive an approximate solution in terms of the piecewise bilinear functions employed earlier. We introduce the function

$$w(x,y) = [(\tfrac{1}{2}\pi)^2 - x^2](\tfrac{1}{2}\pi + y)\frac{1}{\pi},$$

which satisfies the boundary conditions, and then we define a Galerkin approximation of the form

$$U(x,y) = \sum_{i,j=1}^{T} \alpha_{ij}\varphi_{ij}(x,y) + w(x,y),$$

by means of (3.21). The results of the computation, using the various grids described earlier, are given in Table 2. Figure 10 illustrates the position of the points referred to in the table. Although there is not the same degree of symmetry as before, it is still only necessary to consider half of the region. The theoretical solution of equation (3.26) subject to the boundary conditions (3.33) is

$$u(x,y) = \frac{4}{\pi}e^{\frac{1}{2}\pi-y}\sum_{k=1}^{\infty}\left\{\frac{\sin([2k-1]^2+1)^{1/2}y}{\sin([2k-1]^2+1)^{1/2}\tfrac{1}{2}\pi}\right.$$
$$\left.+\frac{\cosh([2k-1]^2+1)^{1/2}y}{\cosh([2k-1]^2+1)^{1/2}\tfrac{1}{2}\pi}\right\}\frac{\cos(2k-1)x}{(-1)^{k-1}(2k-1)^3}.$$

It should be emphasized at this point that *all* the Galerkin methods generalize to systems of equations without any great effort.

Table 2

		Solution		
Point	T = 3	T = 7	T = 15	Exact
1	1.826	1.822	1.821	1.821
2	1.324	1.314	1.311	1.310
3	0.934	0.870	0.859	0.855
4	1.301	1.309	1.310	1.311
5	0.940	0.933	0.931	0.931
6	0.665	0.617	0.608	0.605

56

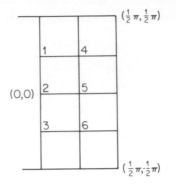

Figure 10

Semi-discrete Galerkin methods

If the Galerkin method is formulated from a variational principle by
introducing the adjoint problem it should be clear that by applying the
technique developed in Section 3.3 it is possible to reduce a dissipative
partial differential equation to a system of ordinary differential
equations. Such methods are of little value for solving boundary value
problems and, as the variational formulation is not valid for initial
value problems, we shall not consider this method in detail. Any reader
interested in this formulation should work through the exercise given
below.

Exercise 6   Let

$$w(x,y) = \sum_{i=1}^{2} \alpha_i(x)\, \varphi_i(y) \tag{3.34a}$$

and

$$w^*(x,y) = \sum_{i=1}^{2} \beta_i(x)\, \varphi_i(y), \tag{3.34b}$$

where φ_1 and φ_2 are piecewise linear functions. Construct the functional
$I(w,w^*)$, corresponding to equation (3.26) subject to the boundary
conditions (3.33). Then with $w(x,y)$ and $w^*(x,y)$ as given in (3.34a) and
(3.34b), find the solution of

$$\delta_{\beta_i} J[\alpha_1,\alpha_2,\beta_1,\beta_2] = 0 \quad (i = 1,2),$$

and hence determine the semi-discrete Galerkin solution to (3.26)
subject to (3.33). Compare your solution with the results given in Table
2.

Alterating Direction Galerkin (A.D.G.) methods for rectangular regions
(Douglas and Dupont, 1971)

When finite element methods are applied to linear problems in one
dimension, the resulting algebraic system has a simple *banded* form,
whereas problems in more than one dimension give rise to *block
banded* matrices in which each block is itself banded. For example in
two dimensions a bilinear approximation often leads to a matrix of the
form:

$$
G = \begin{bmatrix}
D & E & & & & \\
C & D & E & & & \\
& & \cdot & \cdot & \cdot & \\
& & \cdot & \cdot & \cdot & \\
& & & \cdot & \cdot & \cdot \\
& & & C & D & E \\
& & & & E & D
\end{bmatrix}
$$

where C, D and E are tridiagonal submatrices.

There are several algorithms that solve a banded algebraic system
efficiently and cheaply (i.e. maximum accuracy with mimimum time
and storage), but such methods are less efficient and much more expen-
sive when applied to block banded matrices. The object of A.D.G.
procedures in finite element methods is similar to that of A.D.I.
procedures in finite difference methods (Mitchell, 1967), namely to
reduce the algebraic system derived from a multidimensional problem
to a sequence of algebraic systems similar in form to those derived from
one-dimensional problems.

In order to apply A.D.G. methods it is necessary to assume that the
basis functions are of the *tensor product form*, that is

$$\varphi_{ij}(x,y) = \varphi_i(x)\varphi_j(y) \quad (i = 1, \ldots, N_x; j = 1, \ldots, N_y).$$

The matrix $G = \{a(\varphi_{ij}, \varphi_{kl})\}$ can then be factorized such that

$$a(\varphi_{ij},\varphi_{kl}) = a_x(\varphi_i,\varphi_k)b_y(\varphi_j,\varphi_l) + b_x(\varphi_i,\varphi_k)a_y(\varphi_j,\varphi_l)$$
$$(i,k = 1, \ldots, N_x; j,l = 1, \ldots, N_y).$$

If A is an $N \times N$ matrix and B is an $M \times M$ matrix the *matrix tensor
product*, denoted by $A \otimes B$ is the $NM \times NM$ matrix:

$$
\begin{bmatrix}
a_{11}B & \cdots & a_{1N}B \\
\vdots & & \vdots \\
a_{N1}B & \cdots & a_{NN}B
\end{bmatrix}.
$$

The matrix can then be written as

$$G = A_x \otimes B_y + B_x \otimes A_y,$$

if the nodes are ordered by columns. The algebraic system can now be written as

$$\{A_x \otimes B_y + B_x \otimes A_y\}\alpha = b,$$

which is solved by means of the iteration

$$(\lambda_n B_x + A_x) \otimes (\lambda_n B_y + A_y)\alpha^{(n)}$$
$$= (\lambda_n B_x - A_x) \otimes (\lambda_n B_y - A_y)\alpha^{(n-1)} + 2\lambda_n b = \psi^{(n-1)},$$

as a two-stage procedure,

$$(\lambda_n B_x + A_x) \otimes I_{N_y}\alpha^{(n^*)} = \psi^{(n-1)} \qquad (3.35a)$$

and

$$I_{N_x} \otimes (\lambda_n B_y + A_y)\alpha^{(n)} = \alpha^{(n^*)}. \qquad (3.35b)$$

It is possible to split these equations such that if

$$\alpha_{p,C} = (\alpha_{p1}, \ldots, \alpha_{pN_y})^T \qquad (p = 1, \ldots, N_x)$$

is a column of values on the grid and

$$\alpha_{p,R} = (\alpha_{1p}, \ldots, \alpha_{N_x p})^T \qquad (p = 1, \ldots, N_y)$$

is a row of values on the grid, (3.35a) becomes

$$(\lambda_n B_x + A_x)\alpha_{p,R}^{(n^*)} = \psi_{p,R}^{(n-1)} \qquad (p = 1, \ldots, N_y)$$

and (3.35b) becomes

$$(\lambda_n B_y + A_y)\alpha_{p,C}^{(n)} = \alpha_{p,C}^{(n^*)} \qquad (p = 1, \ldots, N_x).$$

Douglas and Dupont have shown that, with a suitable choice for the sequence of iteration parameters $\{\lambda_n\}$, A.D.G. is a rapidly convergent iterative procedure.

Method of collocation

The method of collocation is similar in many respects to Galerkin's method. It involves selecting the coefficients α_i $(i = 1, \ldots, N)$ in the approximation

$$U = \sum_{i=1}^{N} \alpha_i \varphi_i,$$

such that the differential equation is satisfied exactly at certain specified points. It has been shown (de Boor and Swartz, 1973; Lucas and Reddien, 1972; also Ahlberg and Ito, 1975) that for ordinary differential equations if the collocation points are chosen correctly, then the method is similar in accuracy to Galerkin's method with the same set of basis functions φ_i $(i = 1, \ldots, N)$. If for example, the basis functions are Hermite piecewise polynomials of degree $2r-1$, then the collocation points in each subinterval $[x_i, x_{i+1}]$ are taken as the zeroes of the Legendre polynomial

$$P_r \left(\frac{2x - x_{i+1} - x_i}{x_{i+1} - x_i} \right).$$

The advantages of the method of collocation are:

(i) There are no inner products to integrate as in Galerkin and Ritz approximations.
(ii) The resultant algebraic equations have fewer terms than the corresponding equations for Galerkin approximations.

The main disadvantage of collocation is that it is necessary to use basis functions of degree (at least) $2k$ for a differential equation of order $2k$.

Methods using collocation have also been devised (Douglas and Dupont, 1973) for evolutionary problems. An example involving collocation is given in Section 7.4(B).

3.5 PROJECTION METHODS

Earlier in this chapter the equivalence of a variational equation

$$\delta_v I(v) = 0 \qquad\qquad\qquad (3.36)$$

and a differential equation

$$Au = f \qquad\qquad\qquad (3.37)$$

is used to obtain a Ritz approximate solution of the form

$$U = \sum_{i-1}^{N} \alpha_i \varphi_i.$$

The parameters α_i $(i = 1, \ldots, N)$ are determined from the variational equations

$$\delta_{\alpha_i} I \left(\sum_{j=1}^{N} \alpha_j \varphi_j \right) = 0 \quad (i = 1, \ldots, N)$$

or more explicitly

$$\sum_{j=1}^{N} a(\varphi_i, \varphi_j) \alpha_j = (\varphi_i, f) \quad (i = 1, \ldots, N). \qquad\qquad (3.38)$$

In Section 1.3 it is shown that, given a Hilbert space $\mathcal{H}$ with inner-product $[,]$ and any N-dimensional subspace $K_N \subset \mathcal{H}$, then for any

$q \in \mathscr{H}$ the orthogonal projection

$$\tilde{q} = \sum_{i=1}^{N} \alpha_i \varphi_i \qquad (3.39)$$

is such that

$$\sum_{j=1}^{N} [\varphi_i, \varphi_j] \alpha_j = [\varphi_i, q] \quad (i = 1, \ldots, N), \qquad (3.40)$$

where φ_i $(i = 1, \ldots, N)$ form a basis for K_N. It follows that $\tilde{q}$ is the unique best approximation as in Section 1.2. It is now shown that not only do (3.38) and (3.40) appear similar, but that the approximations possess similar properties.

Let $\mathscr{H}$ be the space of admissible functions corresponding to the differential equation

$$Au = f$$

subject to a given set of boundary conditions. Then the *energy space* $\mathscr{H}_A$ of the operator A, for the particular problem, is the subspace of $\mathscr{H}$ containing all admissible functions $u(\mathbf{x})$ and $v(\mathbf{x})$ such that $a(u,v)$, as defined in Section 3.4, is a bounded functional. The inner product for the energy space $\mathscr{H}_A$, is defined as

$$(u,v)_A = a(u,v)$$

and the norm — which is called the *energy norm* — as

$$\| u \|_A^2 = (u,u)_A .$$

In this definition it is assumed that if it is possible to transform the inner product $(,)_A$ using integration by parts, then this form is used in the definition of the energy space. If this procedure is adopted, the energy space is complete and is a Hilbert space.

Theorem 3.1   *If the linear operator A is positive and self-adjoint, the Ritz approximate solution of the differential equation*

$$Au = f$$

is the orthogonal projection of the theoretical solution onto the approximate subspace of the energy space. Thus the Ritz approximation is the best approximation in terms of the energy space.

Proof.   The proof is a direct consequence of the definition of the Ritz approximation given in Section 3.1. The best approximation

$$U = \sum_{i=1}^{N} \alpha_i \varphi_i ,$$

of the theoretical solution u_0, in terms of the energy norm, is $U \in K_N$

such that

$$(U - u_0, \varphi_i)_A = 0 \quad (i = 1, \ldots, N),$$

that is

$$a(U - u_0, \varphi_i) = 0 \quad (i = 1, \ldots, N). \tag{3.41}$$

If it is assumed that the solution is unique it is possible to rewrite (3.41) as (3.38). Hence the required result follows provided that a is a valid norm. It can be shown (Mikhlin, 1964, pp. 74–78) that the bilinear form a is a norm if and only if (3.36) is equivalent to (3.37), hence the proof is complete.

Thus for example, the Ritz approximate solution of (3.24) is the best approximation to the theoretical solution, in terms of the *Dirichlet semi-norm*

$$| u |_{1,R} = \left\{ \iint_R \left\{ \left(\frac{\partial u}{\partial x}\right)^2 + \left(\frac{\partial u}{\partial y}\right)^2 \right\} dx dy \right\}^{1/2}.$$

Exercise 7 (i) Show that for $(,)_A$ to be an inner product and for $\|.\|_A$ to be a norm the operator A must be positive definite and self-adjoint.

(ii) Show that $a(u,v)$ is bounded if and only if both $\| u \|_A$ and $\| v \|_A$ are bounded. Hence confirm that u is an element of the energy space if and only if $\| u \|_A$ is bounded.

Exercise 8 Show that the functional $I(v)$, corresponding to the differential equation

$$Au = f$$

in R, subject to $u = 0$ on the boundary ∂R, can be rewritten as

$$I(v) = \| v - u_0 \|_A^2 - \| u_0 \|_A^2 \tag{3.42}$$

or

$$I(v) = \| v \|_A^2 - 2(v, u_0)_A,$$

where u_0 is the solution of the differential equation.

Exercise 9 For which group of differential equations is the Ritz approximation best in terms of the Sobolev norm

$$\| u \|_{1,R} = \left\{ \iint_R \left\{ \left(\frac{\partial u}{\partial x}\right)^2 + \left(\frac{\partial u}{\partial x}\right)^2 + u^2 \right\} dx dy \right\}^{1/2}.$$

Method of least squares

The Ritz method gives us the best approximation, to the solution of a linear differential equation

$$Au = f,$$

in terms of the energy norm, if and only if the operator A is positive definite and self-adjoint. Although the Galerkin method has extended the class of problems for which we can find approximate solutions, it does not lead to the same type of best approximation. To derive 'best approximations' for non-self-adjoint problems it is necessary to reformulate the approximation procedure and introduce the so-called method of least squares.

We recall that the Ritz approximation was the solution of

$$\underset{U \in K_N}{\text{minimum}} \, \| U - u_0 \|_A^2$$

where u_0 is the unknown solution. It is theoretically possible to replace the energy norm $\| . \|_A$ by any other norm, *provided* that we only require Au_0 and not u_0 itself. Thus we can define an approximation U as the solution of

$$\underset{U \in K_N}{\text{minimum}} \, \| AU - Au_0 \|^2$$

or

$$\underset{U \in K_N}{\text{minimum}} \, \| AU - f \|^2 \, ;$$

this is the basis of the method of least squares (Bramble and Schatz, 1970; 1971). If we apply the calculus of variations to determine the conditions necessary for a stationary value of such a functional, we obtain an Euler—Lagrange equation involving the operator $A*A$, that is, an equation of higher order than the one given. For this reason it follows that we must take care to ensure that the solutions are the same, particularly when inhomogeneous Dirichlet boundary conditions are specified; thus unless we impose additional conditions on the approximating functions, we must modify the functional (Section 3.2). If we are given

$$u = g$$

on ∂R, a least-squares approximation could minimize

$$\iint_R (AU - f)^2 \, \mathrm{d}x + \lambda \int_{\partial R} (U - g)^2 \, \mathrm{d}\sigma,$$

for a suitable choice of λ.

It is beyond the scope of this book to cover the method of least squares in much greater detail, but it is mentioned again in Section 5.4(D). Numerical experiments involving the method of least squares can be found in Chapter 7.

Chapter 4

Basis Functions

In Section 1.1, elementary basis functions were constructed for rectangular and polygonal regions, where the former was divided up into a number of rectangular elements and the latter into a number of triangular elements. In this chapter, a study is made of the construction of basis functions for a variety of element shapes in two and three dimensions.

4.1 THE TRIANGLE

(A) Lagrange interpolation

The triangle or two-dimensional simplex is probably the most widely used finite element. One reason for this is that arbitrary regions in two dimensions can be approximated by polygons, which can always be divided up into a finite number of triangles. In addition, the complete mth-order polynomial

$$\Pi_m(x,y) = \sum_{k+l=0}^{m} \alpha_{kl} x^k y^l \qquad (4.1)$$

can be used to interpolate a function, say $U(x,y)$, at $\frac{1}{2}(m+1)(m+2)$ symmetrically placed nodes in a triangle. The first three cases of this general representation for the triangle $P_1 P_2 P_3$, with the coordinates of the vertices being (x_1,y_1), (x_2,y_2) and (x_3,y_3) respectively, are:

(1) *The linear case* ($m = 1$). Here the polynomial is

$$\Pi_1(x,y) = \alpha_1 + \alpha_2 x + \alpha_3 y$$

$$= \sum_{j=1}^{3} U_j p_j^{(1)}(x,y),$$

where U_j ($j = 1,2,3$) are the values of $U(x,y)$ at the vertices P_j and

$$p_j^{(1)}(x,y) = \frac{1}{C_{jkl}}(\tau_{kl} + \eta_{kl}x - \xi_{kl}y)$$

$$= \frac{D_{kl}}{C_{jkl}}, \qquad (4.2)$$

where

$$\tau_{kl} = x_k y_l - y_l x_k, \quad \xi_{kl} = x_k - x_l, \quad \eta_{kl} = y_k - y_l,$$

and

$$D_{kl} = \det \begin{bmatrix} 1 & x & y \\ 1 & x_k & y_k \\ 1 & x_l & y_l \end{bmatrix},$$

with (j,k,l) any permutation of $(1,2,3)$, and the modulus of

$$C_{jkl} = \det \begin{bmatrix} 1 & x_j & y_j \\ 1 & x_k & y_k \\ 1 & x_l & y_l \end{bmatrix}$$

is twice the area of the triangle $P_1 P_2 P_3$. It is easily seen that

$$p_j^{(1)}(x_k, y_k) = \begin{cases} 1 & (j = k) \\ 0 & (j \neq k) \end{cases} \quad (1 \leqslant j, k \leqslant 3)$$

(2) *The quadratic case* ($m = 2$). The polynomial is now

$$\Pi_2(x,y) = \sum_{j=1}^{6} U_j p_j^{(2)}(x,y), \tag{4.3}$$

where U_j ($j = 1, \ldots, 6$) are the values of $U(x,y)$ at the vertices P_j ($j = 1,2,3$) together with the values at the mid-points P_j ($j = 4,5,6$) of the sides $P_1 P_2$, $P_2 P_3$ and $P_3 P_1$ respectively. The functions $p_j^{(2)}(x,y)$ ($j = 1,2, \ldots, 6$) are given by

$$p_1^{(2)}(x,y) = p_1^{(1)}(2p_1^{(1)} - 1),$$

with $p_2^{(2)}(x,y)$ and $p_3^{(2)}(x,y)$ similarly, and

$$p_4^{(2)}(x,y) = 4p_1^{(1)} p_2^{(1)}$$

with $p_5^{(2)}(x,y)$ and $p_6^{(2)}(x,y)$ similarly. Again it follows that

$$p_j^{(2)}(x_k, y_k) = \begin{cases} 1 & (j = k) \\ 0 & (j \neq k) \end{cases} \quad (1 \leqslant j, k \leqslant 6).$$

It is particularly satisfactory that the basis functions $p_j^{(2)}(x,y)$ ($j = 1, \ldots, 6$) can be expressed in terms of the basis function $p_j^{(1)}(x,y)$ ($j = 1,2,3$). This is true for nearly all the basis functions we shall consider and therefore in order to simplify the formulae we shall denote $p_j^{(1)}$ simply by p_j ($j = 1,2,3$).

(3) *The cubic case* ($m = 3$). The polynomial is

$$\Pi_3(x,y) = \sum_{j=1}^{10} U_j p_j^{(3)}(x,y), \qquad (4.4)$$

where U_j ($j = 1,2,3$) are the values of $U(x,y)$ at the vertices P_1, P_2, P_3, U_j ($j = 4,5, \ldots , 9$) are the values at the points of trisection of the sides, and U_{10} is the value of $U(x,y)$ at the centroid of the triangle. The basis functions are given by

$$p_1^{(3)}(x,y) = \tfrac{1}{2}p_1(3p_1 - 1)(3p_1 - 2),$$

with $p_2^{(3)}(x,y)$ and $p_3^{(3)}(x,y)$ similarly,

$$p_4^{(3)}(x,y) = \tfrac{9}{2}p_1 p_2(3p_1 - 1),$$
$$p_5^{(3)}(x,y) = \tfrac{9}{2}p_1 p_2(3p_2 - 1),$$

with $p_6^{(3)}, \ldots , p_9^{(3)}$ similarly, and

$$p_{10}^{(3)}(x,y) = 27 p_1 p_2 p_3.$$

The tenth parameter can be eliminated by using the linear relation

$$U_{10} = \frac{1}{4} \sum_{j=4}^{9} U_j - \frac{1}{6} \sum_{j=1}^{3} U_j$$

to yield a function that will still interpolate quadratics exactly (Ciarlet and Raviart, 1972a); this is an example of the so-called *elimination of internal parameters*. The triangles for the cases $m = 1,2,3$, are shown in Figure 11.

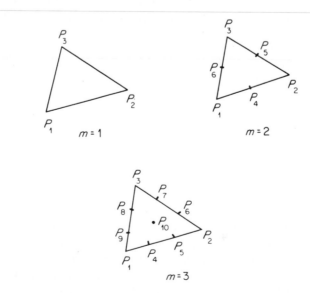

Figure 11

Exercise 1   Verify that

$$U(x,y) = \sum_{j=1}^{9} U_j \bar{p}_j^{(3)}(x,y)$$

interpolates quadratic polynomials exactly if

$$\bar{p}_j^{(3)} = p_j^{(3)} + \alpha_j p_{10}^{(3)} \quad (j = 1, \ldots, 9),$$

with

$$\alpha_j = \begin{cases} -\tfrac{1}{6} & (j = 1,2,3) \\ \tfrac{1}{4} & (j = 4, \ldots, 9). \end{cases}$$

Exercise 2   Show that

$$p_j^{(3)}(x_k,y_k) = \begin{cases} 1 & (k = j) \\ 0 & (k \neq j) \end{cases} \quad (1 \leqslant j, k \leqslant 10).$$

Exercise 3   Express

$$\bar{p}_j^{(3)}(x,y) \quad (j = 5,6,7,8,9)$$

in terms of

$$p_k(x,y) \quad (k = 1,2,3).$$

We now turn to the general case where the complete mth-order polynomial is given by (4.1). This polynomial has $\tfrac{1}{2}(m+1)(m+2)$ coefficients which can be chosen so that the polynomial interpolates $U(x,y)$ at the $\tfrac{1}{2}(m+1)(m+2)$ symmetrically placed points on the triangle $P_1 P_2 P_3$ whose coordinates are given by

$$\sum_{l=1}^{3} \frac{\beta_l x_l}{m}, \quad \sum_{l=1}^{3} \frac{\beta_l y_l}{m}, \tag{4.5}$$

where β_1, β_2 and β_3 are integers satisfying $0 \leqslant \beta_k \leqslant m$ ($k = 1,2,3$) and $\beta_1 + \beta_2 + \beta_3 = m$. These points include the three vertices of the triangle $P_1 P_2 P_3$. The remaining points are obtained geometrically by dividing each side of the triangle into m equal parts and joining the points of subdivision by lines parallel to the sides of the triangle. This subdivides the triangle into m^2 congruent triangles whose vertices are the $\tfrac{1}{2}(m+1)(m+2)$ points described by (4.5). If U_j denotes the value of $U(x,y)$ at a point given by (4.5), the interpolating polynomial of degree m can be expressed as

$$U(x,y) = \sum_{j=1}^{\tfrac{1}{2}(m+1)(m+2)} U_j p_j^{(m)}(x,y), \tag{4.6}$$

where the summation is over all $\tfrac{1}{2}(m+1)(m+2)$ points, and $p_j^{(m)}(x,y)$ is a polynomial basis function of degree m taking the value unity at the point associated with the triple $(\beta_1,\beta_2,\beta_3)$ and the value zero at every other point. The formula (4.6) is a *Lagrange* type interpolating formula.

The standard triangle

It should be noted from (4.2) that

(i) $\sum\limits_{j=1}^{3} p_j(x,y) = 1$

(ii) the linear equations $p_j(x,y) = 0$ ($j = 1,2,3$) represent the triangle sides P_2P_3, P_3P_1 and P_1P_2 respectively, and

(iii) $p_j(x,y) = 1$ ($j = 1,2,3$) at the vertices P_1, P_2 and P_3 respectively.

Alternatively, the triangle $P_1P_2P_3$ in the (x,y) plane is transformed into the *standard triangle* $\Pi_1\Pi_2\Pi_3$ in the (p_1,p_2) plane by the transformation formulae (4.2), where $\Pi_1 = (1,0)$, $\Pi_2 = (0,1)$ and $\Pi_3 = (0,0)$ (see Figure 12). The inverse transformation from the (p_1,p_2) plane to the (x,y) plane is given by

$$x = x_3 + \xi_{13}p_1 + \xi_{23}p_2$$

and (4.7)

$$y = y_3 + \eta_{13}p_1 + \eta_{23}p_2.$$

Since all triangles in a triangular network in the (x,y) plane can be transformed into this standard triangle, it is very convenient to work in terms of the standard triangle, and at an appropriate point to transfer the result back to a particular triangle in the (x,y) plane through the linear transformation (4.7). This procedure will be used repeatedly when triangular elements are involved both in this chapter and in Chapter 5.

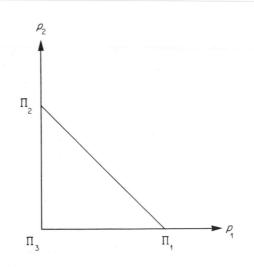

Figure 12

Geometrical method of constructing basis functions

A simple geometrical procedure for constructing basis functions obtained from Lagrange interpolation is now illustrated with respect to the quadratic and cubic cases of the triangle. In the former, for example, the basis function $p_1^{(2)}(x,y)$ must take the value zero at the nodes P_j ($j = 2,3, \ldots , 6$) and the value unity at the node P_1. The line $p_1 = 0$ passes through the points P_2, P_3 and P_5, and the line $p_1 = \frac{1}{2}$ passes through the points P_4 and P_6, and so the basis function $p_1^{(2)}(x,y) = p_1(2p_1 - 1)$ is obtained. The functions $p_2^{(2)}(x,y)$ and $p_3^{(2)}(x,y)$ are similarly obtained. The basis function at a mid-side node, say P_4, is obtained from the lines $p_1 = 0$ and $p_2 = 0$, which pass through the points P_2, P_3, P_5, and P_1, P_3, P_6 respectively. The required basis function, normalized at P_4, is $p_4^{(2)}(x,y) = 4p_1 p_2$. The functions $p_5^{(2)}(x,y)$ and $p_6^{(2)}(x,y)$ are similarly obtained. The cubic case follows in an analogous manner.

C^0 approximating functions

In the general case, the function interpolates the values of $U(x,y)$ at $\frac{1}{2}(m + 1)(m + 2)$ points on the triangle. On a side of the triangle, this function reduces to a polynomial of degree m in the variable s which is measured along the side of the triangle. The polynomial interpolates $U(x,y)$ at $(m + 1)$ points on the side of the triangle, and so is unique. Also in a triangular network, each side (internal) belongs to two triangles. If the function in each triangle interpolates at $\frac{1}{2}(m + 1)(m + 2)$ symmetrically placed points, it will reduce to the unique polynomial of degree m in s on the common side. This means that the interpolating function over the complete triangular network is continuous along internal sides of the network and so has C^0 continuity over the polygonal region.

(B) Hermite interpolation

As an alternative to interpolating the function $U(x,y)$ at a large number of points symmetrically placed on the triangle, it is possible to interpolate $U(x,y)$ and some of its derivatives at a smaller number of points.

A class of polynomials suited to this particular task consists of the complete polynomials $G_\nu(x,y)$ of odd degree $2\nu + 1$ ($\nu = 1,2,3, \ldots$), which are determined by the values

$$D^i G_\nu(P_j) \quad (|i| \leqslant \nu; j = 1,2,3)$$

and

$$D^i G_\nu(P_4) \quad (|i| \leqslant \nu - 1).$$

(4.8)

Here P_1, P_2, P_3 are the vertices of the triangle, and P_4 is the centroid;

$i = (i_1, i_2)$ where i_1, i_2 are nonnegative integers, $|i| = i_1 + i_2$ and

$$D^i G = \frac{\partial^{|i|} G}{\partial x^{i_1} \partial y^{i_2}}.$$

This is an example of the *multi-index notation* for derivatives. The first two cases of this general representation are:

(1) *The cubic case* $(\nu = 1)$. Here the complete cubic polynomial has ten coefficients which are uniquely determined by matching the function values and the first-order partial derivatives at the vertices and the function value at the centroid. Thus in this case we may write the polynomial $\Pi_3(x,y)$, which is given in (4.4), in the form

$$G_1(x,y) = \sum_{j=1}^{4} U_j q_j^{(3)}(x,y) + \sum_{j=1}^{3} \left[\left(\frac{\partial U}{\partial x} \right)_j r_j^{(3)}(x,y) \right.$$
$$\left. + \left(\frac{\partial U}{\partial y} \right)_j s_j^{(3)}(x,y) \right], \quad (4.9)$$

where

$$q_j^{(3)}(x,y) = p_j(3p_j - 2p_j^2 - 7p_k p_l);$$

with (j,k,l) any cyclic permutation of $(1,2,3)$ and

$$q_4^{(3)}(x,y) = 27 p_1 p_2 p_3.$$

Also

$$r_j^{(3)}(x,y) = p_j[\xi_{jk} p_k(p_l - p_j) + \xi_{jl} p_l(p_k - p_j)]$$

and $s_j^{(3)}(x,y)$ is obtained from $r_j^{(3)}(x,y)$ by replacing ξ by η. This element is in common use in the finite element method.

(2) *The quintic case* $(\nu = 2)$. This time the complete quintic polynomial has twenty-one coefficients which are uniquely determined by matching the function values and the first- and second-order partial derivatives at the vertices, and the function value and the first-order partial derivatives at the centroid. This element is of little practical use and will not be considered further.

Exercise 4 Show that on a side of the triangle the function given by (4.9) reduces to a polynomial of degree 3 in the variable s which is measured along the side of the triangle. Show also that this polynomial is uniquely determined by the values of the function and its first-order partial derivatives at the two vertices which are the end-points of the side. Hence show that this element gives an interpolating function which has C^0 continuity over a triangular network.

As for cubic Lagrange interpolation it is possible to eliminate the value of $U(x,y)$ at the centroid and still interpolate quadratic polynomials exactly. In this case the replacement is

$$U_4 = \frac{1}{3} \sum_{j=1}^{3} U_j + \frac{1}{18} \Sigma^{(1)} \left\{ \left(\frac{\partial U}{\partial x}\right)_j (\xi_{kj} + \xi_{lj}) + \left(\frac{\partial U}{\partial y}\right)_j (\eta_{kj} + \eta_{lj}) \right\},$$

where $\Sigma^{(1)}$ denotes the summation over (j,k,l) for all cyclic permutations of $(1,2,3)$.

The interpolating polynomial now becomes

$$G_1^*(x,y) = \sum_{j=1}^{3} \left[U_j q_j^{(3)*}(x,y) + \left(\frac{\partial U}{\partial x}\right)_j r_j^{(3)*}(x,y) \right.$$
$$\left. + \left(\frac{\partial U}{\partial y}\right)_j s_j^{(3)*}(x,y) \right], \quad (4.10)$$

where

$$q_j^{(3)*}(x,y) = p_j(3p_j - 2p_j^2 + 2p_k p_l) \quad (4.10a)$$

and

$$r_j^{(3)*}(x,y) = p_j^2(p_l \xi_{lj} + p_k \xi_{kj}) + \tfrac{1}{2}p_j p_k p_l(\xi_{lj} + \xi_{kj}), \quad (4.10b)$$

with (j,k,l) any permutation of $(1,2,3)$. The functions $s_j^{(3)*}$ are obtained from (4.10b) by replacing ξ by η.

Exercise 5   Verify that quadratic functions are interpolated exactly by the reduced cubic interpolant.

Exercise 6   By using the linear transformation formulae (4.7), show that (4.10) becomes

$$G_1^*(p_1,p_2) = \sum_{j=1}^{3} \left[U_j q_j^{(3)*} + \left(\frac{\partial U}{\partial p_1}\right)_j r_j^{(3)*} + \left(\frac{\partial U}{\partial p_2}\right)_j s_j^{(3)*} \right],$$

in the variables of the standard triangle, where

$$\frac{\partial U}{\partial p_j} = \xi_{j3} \frac{\partial U}{\partial x} + \eta_{j3} \frac{\partial U}{\partial y} \quad (j = 1,2)$$

and the functions $q_j^{(3)*}$, $r_j^{(3)*}$ and $s_j^{(3)*}$ $(j = 1,2,3)$ are given by

$$q_j^{(3)*} = p_j^2(3 - 2p_j) + 2p_1 p_2 p_3,$$

$$r_j^{(3)*} = \begin{cases} p_1^2(p_1 - 1) - p_1 p_2 p_3 & (j = 1) \\ p_j^2 p_1 + \tfrac{1}{2}p_1 p_2 p_3 & (j = 2,3) \end{cases}$$

and

$$s_j^{(3)*} = \begin{cases} p_2^2(p_2 - 1) - p_1 p_2 p_3 & (j = 2) \\ p_j^2 p_2 + \tfrac{1}{2}p_1 p_2 p_3 & (j = 1,3). \end{cases}$$

You may use the results

$$\xi_{kl} \eta_{lj} - \xi_{lj} \eta_{kl} = C_{jkl},$$

where (j,k,l) is any permutation of $(1,2,3)$.

Exercise 7   In the special case of $P_1 = (1,0)$, $P_2 = (0,1)$, $P_3 = (0,0)$, find $G_1^*(x,y)$ from equation (4.10). Hence show that the normal derivatives of $G_1^*(x,y)$ to the sides are given by

$$\left(\frac{\partial G_1^*}{\partial x}\right)_{P_2 P_3} = y(1-y)\left[2U_1 - \left(\frac{\partial U}{\partial x}\right)_1 + \frac{1}{2}\left(\frac{\partial U}{\partial y}\right)_1 + 2U_2\right.$$

$$-\frac{1}{2}\left(\frac{\partial U}{\partial x}\right)_2 - \left(\frac{\partial U}{\partial y}\right)_2 - 4U_3 - \frac{1}{2}\left(\frac{\partial U}{\partial x}\right)_3 - \frac{3}{2}\left(\frac{\partial U}{\partial y}\right)_3\right]$$

$$+ y\left(\frac{\partial U}{\partial x}\right)_2 + (1-y)\left(\frac{\partial U}{\partial x}\right)_3,$$

$$\left(\frac{\partial G_1^*}{\partial y}\right)_{P_3 P_1} = x(1-x)\left[2U_1 - \left(\frac{\partial U}{\partial x}\right)_1 - \frac{1}{2}\left(\frac{\partial U}{\partial y}\right)_1 + 2U_2\right.$$

$$+\frac{1}{2}\left(\frac{\partial U}{\partial x}\right)_2 - \left(\frac{\partial U}{\partial y}\right)_2 - 4U_3 - \frac{3}{2}\left(\frac{\partial U}{\partial x}\right)_3 - \frac{1}{2}\left(\frac{\partial U}{\partial y}\right)_3\right]$$

$$+ x\left(\frac{\partial U}{\partial y}\right)_1 + (1-x)\left(\frac{\partial U}{\partial y}\right)_3$$

and

$$\left(\frac{\partial G_1^*}{\partial X}\right)_{P_1 P_2} = \frac{1}{4}(1-Y)^2\left[U_1 - \frac{1}{2}\left(\frac{\partial U}{\partial X}\right)_1 + \frac{1}{2}\left(\frac{\partial U}{\partial Y}\right)_1 + U_2 - \frac{1}{2}\left(\frac{\partial U}{\partial X}\right)_2\right.$$

$$-\frac{1}{2}\left(\frac{\partial U}{\partial Y}\right)_2 - 2U_3 - \left(\frac{\partial U}{\partial X}\right)_3\right] + \frac{1}{2}(1-Y)\left(\frac{\partial U}{\partial X}\right)_1$$

$$+ \frac{1}{2}(1+Y)\left(\frac{\partial U}{\partial X}\right)_2,$$

where

$$X = x+y, \quad Y = y-x, \quad \frac{\partial U}{\partial X} = \frac{1}{2}\left(\frac{\partial U}{\partial x} + \frac{\partial U}{\partial y}\right)$$

and

$$\frac{\partial U}{\partial Y} = \frac{1}{2}\left(-\frac{\partial U}{\partial x} + \frac{\partial U}{\partial y}\right).$$

Tricubic interpolation

Birkhoff (1971) introduced a triangular element which involves the twelve-parameter family of all quartic polynomials which are cubic along any parallel to any side of a triangle. With respect to the standard triangle, such a family is

$$U(p,q) = \sum_{j+k\leqslant 4} \alpha_{jk}\, p_1^j p_2^k \qquad (4.11)$$

with

$$\alpha_{31} + \alpha_{13} = \alpha_{22}.$$

The polynomial (4.11) is called a *tricubic* polynomial. This polynomial uniquely interpolates values of U, $\partial U/\partial p_1$, $\partial U/\partial p_2$ and $\partial^2 U/\partial r\partial s$ at each vertex, where $\partial^2 U/\partial r\partial s$ is a cross-derivative determined at each vertex with r and s in directions parallel to the adjacent sides. The unique interpolating function takes the form

$$\sum_{j=1}^{3} \left[\left\{ \sum_{|i|\leqslant 1} D^i U_j\, \varphi_j^i(p_1,p_2) \right\} + \left(\frac{\partial^2 U}{\partial r\partial s} \right)_j \hat{\varphi}_j(p_1,p_2) \right], \qquad (4.12)$$

where the suffix denotes the value of the quantity at the vertex Π_j of the standard triangle (Figure 12). The coefficients are given by

$$\varphi_j^{(0,0)} = p_j^2 (3 - 2p_j + 6p_k p_l),$$

$$\varphi_j^{(1,0)} = \begin{cases} p_1^2(p_1 - 1 - 4p_2 p_3) & (j=1) \\ p_2^2 p_1(1 + 2p_3) & (j=2) \\ p_3^2 p_1(1 + 2p_2) & (j=3), \end{cases}$$

$$\varphi_j^{(0,1)} = \begin{cases} p_1^2 p_2(1 + 2p_3) & (j=1) \\ p_2^2 p_2(p_2 - 1 - 4p_1 p_3) & (j=2) \\ p_3^2 p_2(1 + 2p_1) & (j=3) \end{cases}$$

and

$$\hat{\varphi}_j = 2p_j^2 p_k p_l,$$

where (j,k,l) is any permutation of $(1,2,3)$. Note that D^i in this case represents derivatives in the (p_1,p_2) plane. The cross-derivatives $(\partial^2 U/\partial r\partial s)_j$ $(j=1,2,3)$ are given by $-\partial^2 U/\partial p_1\partial Q$, $\partial^2 U/\partial p_2\partial Q$ and $\partial^2 U/\partial p_1\partial p_2$ respectively, where $Q = p_2 - p_1$. The unique tricubic interpolating polynomial in a particular triangle in the (x,y) plane is obtained from (4.12) by using the linear transformation formulae (4.2).

Exercise 8   Show that the tricubic polynomials produce a C^0 approximating function over a network of triangular elements.

(C) C^1 Approximating functions

A triangular element is now introduced which involves the complete family of quintic polynomials. In the plane of the standard triangle, the complete quintic is

$$U(p_1,p_2) = \sum_{j+k \leqslant 5} \alpha_{jk} p_1^j p_2^k. \tag{4.13}$$

The coefficients α_{jk} can be determined in terms of

$$\mathbf{D}^i U(P_j) \quad (|\,i\,| \leqslant 2; j = 1,2,3)$$

and

$$\frac{\partial U}{\partial n}(P_j) \quad (j = 4,5,6),$$

where P_j ($j = 1,2,3$) are the vertices and P_j ($j = 4,5,6$) are the side mid-points. The function $U(p_1,p_2)$ given by (4.13) reduces to a quintic in s along each side of the triangle, which is uniquely determined by the parameters at the vertices which provide six boundary conditions, viz. U, $\partial U/\partial s$, $\partial^2 U/\partial s^2$ at the end-points of each side. The normal derivative to each side, $\partial U/\partial n$, where n is p_2, $p_1 + p_2$ and p_1 respectively is a quartic in s, and is uniquely determined by the parameters $\partial U/\partial n$ and $\partial^2 U/\partial n \partial s$ at the end-points of each side, together with $\partial U/\partial n$ at mid-side points. It has thus been shown that complete quintic polynomials produce an approximating function over a network of triangular elements, which has continuity of displacement and gradient over the complete region. Such an interpolating function is said to be C^1 over the region.

In fact the parameters corresponding to the normal derivatives at the mid-side points can be eliminated without destroying the C^1 continuity over the triangular network. This is accomplished by imposing a cubic variation of the normal derivative along each side, which is equivalent to requiring that in (4.13)

$$\alpha_{41} = \alpha_{14}$$

and

$$5\alpha_{50} + \alpha_{32} + \alpha_{23} + 5\alpha_{05} = 0.$$

This reduces (4.13) to an eighteen-parameter family of polynomials, and the unique interpolating function is given by

$$U(p_1,p_2) = \sum_{j=1}^{3} \sum_{|i| \leqslant 2} \mathbf{D}^i U_j \varphi_j^i(p_1,p_2), \tag{4.14}$$

where

$$\varphi_j^{(0,0)} = \begin{cases} p_j^2(10p_j - 15p_j^2 + 6p_j^3 + 15p_k^2 p_l) & (j = 1,2) \\ p_j^2(10p_j - 15p_j^2 + 6p_j^3 + 30p_k p_l(p_k + p_l)) & (j = 3), \end{cases}$$

with (j,k,l) cyclic permutations of $(1,2,3)$,

$$\varphi_1^{(1,0)} = p_1^2(-4p_1 + 7p_1^2 - 3p_1^3 - 1\tfrac{5}{2}p_2^2 p_3),$$

$$\varphi_2^{(1,0)} = p_1 p_2^2(3 - 2p_2 - \tfrac{3}{2}p_1 - \tfrac{3}{2}p_1^2 + \tfrac{3}{2}p_1 p_2),$$

$$\varphi_3^{(1,0)} = p_1 p_3^2(3 - 2p_3 - 3p_1^2 + 6p_1 p_2),$$

$$\varphi_1^{(0,1)} = p_1^2 p_2(3 - 2p_1 - \tfrac{3}{2}p_2 - \tfrac{3}{2}p_2^2 + \tfrac{3}{2}p_1 p_2),$$

$$\varphi_2^{(0,1)} = p_2^2(-4p_2 + 7p_2^2 - 3p_2^3 - 1\tfrac{5}{2}p_1^2 p_3),$$

$$\varphi_3^{(0,1)} = p_2 p_3^2(3 - 2p_3 - 3p_2^2 + 6p_1 p_2),$$

$$\varphi_1^{(1,1)} = p_1^2 p_2(-1 + p_1 + \tfrac{1}{2}p_2 + \tfrac{1}{2}p_2^2 - \tfrac{1}{2}p_1 p_2),$$

$$\varphi_2^{(1,1)} = p_1 p_2^2(-1 + \tfrac{1}{2}p_1 + p_2 + \tfrac{1}{2}p_1^2 - \tfrac{1}{2}p_1 p_2),$$

$$\varphi_3^{(1,1)} = p_1 p_2 p_3^2,$$

$$\varphi_1^{(2,0)} = p_1^2(\tfrac{1}{2}p_1(1 - p_1)^2 + \tfrac{5}{4}p_2^2 p_3),$$

$$\varphi_2^{(2,0)} = \tfrac{1}{4}p_1^2 p_2^2 p_3 + \tfrac{1}{2}p_1^2 p_2^3,$$

$$\varphi_3^{(2,0)} = \tfrac{1}{2}p_1^2 p_3^2(1 - p_1 + 2p_2),$$

$$\varphi_1^{(0,2)} = \tfrac{1}{4}p_1^2 p_2^2 p_3 + \tfrac{1}{2}p_1^3 p_2^2,$$

$$\varphi_2^{(0,2)} = p_2^2(\tfrac{1}{2}p_2(1 - p_2)^2 + \tfrac{5}{4}p_1^2 p_3)$$

and

$$\varphi_3^{(0,2)} = \tfrac{1}{2}p_2^2 p_3^2(1 + 2p_1 - p_2).$$

Corrective functions

An alternative method of producing a C^1 approximating function over a triangular network is to start with the C^0 approximating function given by (4.10) and to add corrective terms which will increase the continuity of the function to C^1. The main properties of these corrective functions are that they must vanish on the perimeter of the triangle and that they must reduce the normal derivative of the function along the sides of the triangle from quadratic to linear form. At the same time, of course, they must not destroy the continuity of the function and slope inside the triangular element.

One set of corrective functions which is in common use (Zienkiewicz, 1967, p. 117) consists of

$$A_j(p_1, p_2, p_3) = \frac{p_j p_k^2 p_l^2}{(1 - p_k)(1 - p_l)}, \tag{4.15}$$

where (j,k,l) is a permutation of $(1,2,3)$. In fact Dupuis and Göel (1970) have shown that C^1 continuity is obtained if the right-hand sides of (4.10a) and (4.10b) are supplemented by

$$\delta q_k = 2\left[-A_j + \left(2 - 3\frac{L_l}{L_k}\cos\theta_j\right)A_k + \left(2 - 3\frac{L_k}{L_l}\cos\theta_j\right)A_l\right]$$

(4.15a)

and

$$\delta r_j = \frac{1}{2}\left[(\xi_{jk} + \xi_{jl})A_j + \left(3\xi_{lj} + 5\xi_{jk} + 6\eta_{lj}\frac{L_l}{L_k}\sin\theta_j\right)A_k\right.$$
$$\left. + \left(3\xi_{kj} + 5\xi_{jl} + 6\eta_{jk}\frac{L_k}{L_l}\sin\theta_j\right)A_l\right],$$

(4.15b)

where (j,k,l) is any cyclic permutation of $(1,2,3)$, θ_j is the angle subtended at the vertex P_j and L_j is the length of the side opposite P_j; δs_j is obtained from (4.15b) by replacing ξ and η by η and ξ respectively.

Exercise 9   In the special case of $P_1 = (1,0)$, $P_2 = (0,1)$ and $P_3 = (0,0)$, show that the supplements become

$$\delta q_3 = -2\delta q_2 = -2\delta q_1 = 4\delta r_1 = 4\delta s_2 = 4\bar{A} + 4\bar{B} - 2\bar{C},$$
$$\delta r_2 = -\delta s_1 = \tfrac{1}{2}\bar{A} - \tfrac{1}{2}\bar{B},$$
$$\delta r_3 = \tfrac{1}{2}\bar{A} + \tfrac{3}{2}\bar{B} - \tfrac{1}{2}\bar{C}$$

and

$$\delta s_3 = \tfrac{3}{2}\bar{A} + \tfrac{1}{2}\bar{B} - \tfrac{1}{2}\bar{C},$$

where

$$\bar{A} = \frac{xy^2(1 - x - y)^2}{(1 - y)(x + y)},$$

$$\bar{B} = \frac{x^2y(1 - x - y)^2}{(1 - x)(x + y)}$$

and

$$\bar{C} = \frac{x^2y^2(1 - x - y)}{(1 - x)(1 - y)}.$$

Hence, using the result of Exercise 7, show that the normal derivatives are linear along each side of the triangle.

Another set of corrective functions obtained by Clough and Tocher (1965) requires each triangle to be divided into three small triangles having the centroid as a common vertex. The corrective functions are given by

$$A_j(p_1,p_2,p_3) = \begin{cases} p_1 p_2 p_3 - \tfrac{1}{6}p_j^2(3 - 5p_j) & \text{in} \quad T_j \\ \tfrac{1}{6}p_k^2(3p_l - p_k) & \text{in} \quad T_k \\ \tfrac{1}{6}p_l^2(3p_k - p_l) & \text{in} \quad T_l, \end{cases} \qquad (4.16)$$

where T_j is the small triangle opposite vertex P_j, and so on, and where (j,k,l) is any permutation of $(1,2,3)$. The supplements δq_j, δr_j and δs_j are then obtained from (4.15a) and (4.15b) with A_j given by (4.16).

Exercise 10   For the triangle given in Exercise 9, find the corrective functions $\bar{A}$, $\bar{B}$ and $\bar{C}$ from (4.16), and show that on the side $x = 0$

$$\frac{\partial \bar{A}}{\partial x} = -6y(1 - y), \qquad \frac{\partial \bar{B}}{\partial x} = \frac{\partial \bar{C}}{\partial x} = 0,$$

on $y = 0$

$$\frac{\partial \bar{B}}{\partial y} = -6x(1 - x), \qquad \frac{\partial \bar{A}}{\partial y} = \frac{\partial \bar{C}}{\partial y} = 0$$

and on $x + y = 1$

$$\frac{\partial \bar{C}}{\partial X} = \tfrac{3}{2}(1 - Y^2), \qquad \frac{\partial \bar{A}}{\partial X} = \frac{\partial \bar{C}}{\partial X} = 0,$$

where $X = x + y$ and $Y = y - x$. Hence show that the normal derivatives of $q_j + \delta q_j$, $r_j + \delta r_j$ and $s_j + \delta s_j$ vary linearly along the sides. (Hint. Use the result of Exercise 7.)

Exercise 11   Assume that the functions $A_j(x,y)$ are of the form

$$\Delta_{jk} = \sum_{l+m+n=3} \beta_{lmn} p_1^l p_2^m p_3^n$$

in the subtriangle T_k. Verify that

(i)   at the vertices

$$D^i A_j = 0 \qquad (|i| \leqslant 1)$$

and

(ii)   on the sides

$$\frac{\partial A_j}{\partial n} = \begin{cases} 0 & (p_j \neq 0) \\ p_k p_l & (p_j = 0) \end{cases}$$

if and only if

(i) $\Delta_{jj} = p_1 p_2 p_3 + p_j^2(\alpha_{jj}p_j + \alpha_{jk}p_k + \alpha_{jl}p_l)$

and

(ii) $\Delta_{jk} = p_j^2(\alpha_{kj}p_j + \alpha_{kk}p_k + \alpha_{kl}p_l),$

for any constants α_{jj} etc., where (j,k,l) is any permutation of $(1,2,3)$.

Then by equating the functions Δ_{jj} etc. and the derivatives along the internal interfaces, verify that A_j given by (4.16) is C^1 continuous over $P_1P_2P_3$.

A final set of corrective functions is given by

$$A_j(p_1,p_2,p_3) = \frac{p_j^2 p_k}{1 - p_l}$$

where (j,k,l) is a cyclic permutation of $(1,2,3)$. Birkhoff and Mansfield (1974) used these to supplement the tricubic polynomial in order to obtain C^1 continuity over a triangular network. This fifteen-parameter family is most easily expressed in terms of

$$U, \frac{\partial U}{\partial p_1}, \frac{\partial U}{\partial p_2}$$

at the vertices and

$$\frac{\partial U}{\partial n}, \frac{\partial^2 U}{\partial s \partial n}$$

at the side mid-points, where $\partial/\partial s$ and $\partial/\partial n$ denote partial derivatives in the directions of the sides and the normals respectively. Irons (1969) used the same corrective functions to supplement the complete fourth-order polynomial and so obtain an eighteen-parameter family which also gives C^1 continuity over a triangular grid.

4.2 THE RECTANGLE

Rectangular type regions, i.e. regions with sides parallel to the x- and y-axes occur in many problems in physics and engineering. Consequently the rectangular element is of considerable importance and in this section basis functions are constructed for it.

Bicubic hermites

In Section 1.1, bilinear functions in x and y were used to construct basis functions, given by (1.6), each of which was identically zero except for a region composed of four rectangular elements. Over the complete rectangular region the piecewise bilinear approximating function gave C^0 continuity. In this section we consider the bicubic polynomials written in the form

$$H_3(x,y) = \sum_{j=0}^{3} \sum_{k=0}^{3} \alpha_{jk} x^j y^k \qquad (4.17)$$

over the unit square $0 \leqslant x, y \leqslant 1$. The coefficients α_{jk} $(0 \leqslant j, k \leqslant 3)$ can be found uniquely in terms of the values of H_3, $\partial H_3 / \partial x$, $\partial H_3 / \partial y$ and $\partial^2 H_3 / \partial x \partial y$ at the four corners such that

$$H_3(x,y) = \sum_{0 \leqslant j, k, l, m \leqslant 1} (D^{(l,m)} U)_{jk} \, \psi_j^{(l)}(x) \psi_k^{(m)}(y), \qquad (4.18)$$

where

$$\psi_0^{(0)}(t) = (1 - t)^2 (1 + 2t),$$
$$\psi_0^{(1)}(t) = (1 - t)^2 \, t,$$
$$\psi_1^{(0)}(t) = t^2 (3 - 2t)$$

and

$$\psi_1^{(1)}(t) = t^2 (t - 1).$$

The subscript jk denotes the value at the corner $x = j$, $y = k$.

It is an easy matter to see how the result (4.18) can be modified to give the required Hermite bicubic interpolating function over any rectangular element of the original rectangular type region. The approximating function over the complete region is then obtained in a manner similar to that used to obtain (1.6). This time there are *four* basis functions corresponding to each node of the rectangular array. For an internal node, i.e. a node not on the boundary of the rectangular type region, each basis function has a support of four rectangular elements. For a node on the boundary, but not at a corner, the support is two rectangular elements, and for the corner nodes one rectangular element. Over the complete rectangular region the piecewise bicubic approximating function gives $C^{1,1}$ continuity.†

Exercise 12 Using (4.18) show that the basis functions for $D^{(l,m)} U$ $(0 \leqslant l, m \leqslant 1)$ at the node $(0,0)$ of a *unit mesh* are given by the *tensor product form*

$$\varphi^{(l,m)}(x,y) = \varphi^{(l)}(x) \varphi^{(m)}(y),$$

where

$$\varphi^{(0)}(t) = \begin{cases} (1 - t)^2 (1 + 2t) & (0 \leqslant t \leqslant 1) \\ (1 + t)^2 (1 - 2t) & (-1 \leqslant t \leqslant 0) \end{cases}$$

and

$$\varphi^{(1)}(t) = \begin{cases} (1 - t)^2 \, t & (0 \leqslant t \leqslant 1) \\ (1 + t)^2 \, t & (-1 \leqslant t \leqslant 0). \end{cases}$$

†$C^{j,k}$ continuity of $u(x,y)$ means continuity of all the derivatives $D^{(l,m)} u$ $(0 \leqslant l \leqslant j, 0 \leqslant m \leqslant k)$.

An interesting rectangular element has been described by Powell (1973). The rectangle is divided up into four triangles by the diagonals and a full quadratic expression in x and y assumed in each triangle. The coefficients in the quadratics can be chosen to permit C^1 continuity over a rectangular grid.

Bicubic splines

It was shown in Section 1.1, that the cubic spline function in one dimension with local support of $4h$ takes the form (1.15). In fact this spline first suggested by Schoenberg (1969) is often written as

$$M(x) = \frac{1}{6h} \delta^4 \left(\frac{x}{h} - j \right)_+^3 , \tag{4.19}$$

where δ is the usual central difference operator and where the constant $1/6h$ is chosen so that

$$\int_{-\infty}^{+\infty} M(x)\mathrm{d}x = 1.$$

The constant ¼ in (1.15) was chosen to make

$$B_j(j) = 1.$$

In rectangular type regions subdivided into rectangular elements, we consider the Schoenberg splines $M(x)$, given by (4.19), and $M(y)$, given by

$$M(y) = \frac{1}{6h} \delta^4 \left(\frac{y}{h} - k \right)_+^3 .$$

The tensor product of $M(x)$ and $M(y)$ gives rise to a function with a support of sixteen rectangular elements. This is the basis function for cubic splines at the node $x = jh$, $y = kh$, provided the latter is neither on the boundary of the region, nor adjacent to the boundary. For such nodes special basis functions have to be constructed, unless of course the problem being solved has natural boundary conditions (Chapter 3, p. 44). The overall approximating function in this case has $C^{2,2}$ continuity.

It is perhaps worth mentioning that in the unlikely event of $C^{4,4}$ continuity being required of the approximating function on a rectangular region the Schoenberg quintic splines

$$M(x) = \frac{1}{5!h} \delta^6 \left(\frac{x}{h} - j \right)_+^5$$

can be used in a manner similar to the cubic splines of the last section.

The tensor products have a support of thirty-six rectangular elements, which makes the biquintic splines rather difficult to handle.

In concluding this short section on the rectangular element it is worth pointing out that whereas the size of the support of a spline increases with the order, the support of the Hermite function remains constant at four elements, irrespective of the order of the Hermite function, but the spline of course has continuity $C^{2(\nu-1),2(\nu-1)}$ as against $C^{\nu-1,\nu-1}$ for the Hermite function with polynomials of degree $2\nu - 1$.

4.3 THE QUADRILATERAL

It might be thought that quadrilaterals are better mesh units than triangles because the overall grid is simplified. For example, a triangular network can always be simplified by combining the triangles in pairs to form quadrilaterals. Unfortunately, however, it is impossible to find a polynomial in x and y which reduces to an arbitrary linear form along the four sides of a general quadrilateral, and so it is not obvious how one can construct a piecewise function in x and y which has C^0 continuity over a quadrilateral network.

Lemma 4.1   *Let* $\mathscr{P}_1$, $\mathscr{P}_2$, $\mathscr{P}_3$ *and* $\mathscr{P}_4$ *be points in* three-dimensional *space such that* $\mathscr{P}_j = (x_j, y_j, z_j)$ $(j = 1,2,3,4)$. *Then the plane passing through the three points* $\mathscr{P}_j$, $\mathscr{P}_k$ *and* $\mathscr{P}_l$ $(j \neq k \neq l)$ *is*

$$\Pi_{jkl} = 0,$$

where

$$\Pi_{jkl} = -zC_{jkl} + z_j D_{kl} - z_k D_{jl} + z_l D_{jk}$$

with C_{jkl}, D_{kl}, *etc. defined as in Section 4.1. In addition the surface*

$$\alpha\Pi_{klm}\Pi_{jkm} - \beta\Pi_{jlm}\Pi_{jkl} = 0, \tag{4.20}$$

where (j,k,l,m) *is any permutation of* $(1,2,3,4)$, *passes through the four points* $\mathscr{P}_1$, $\mathscr{P}_2$, $\mathscr{P}_3$ *and* $\mathscr{P}_4$, *and contains the lines* $\mathscr{P}_1 \mathscr{P}_2$, $\mathscr{P}_2 \mathscr{P}_3$, $\mathscr{P}_3 \mathscr{P}_4$ *and* $\mathscr{P}_4 \mathscr{P}_1$ *for any values of* α *and* β.

From this lemma, it is clear that surfaces such as (4.20) that pass through points $\mathscr{P}_j = (x_j, y_j, f_j)$ $(j = 1,2,3,4)$, can be used to define functions $f(x,y)$ in the quadrilateral $P_1 P_2 P_3 P_4$, where $P_j = (x_j, y_j)$, such that (i) $f(x_j, y_j) = f_j$ and (ii) $f(x,y)$ varies linearly along the sides $P_1 P_2$, $P_2 P_3$, $P_3 P_4$ and $P_4 P_1$ of the quadrilateral.

Thus it is possible to take $\mathscr{P}_j = (x_j, y_j, 1)$ and $\mathscr{P}_k = (x_k, y_k, 0)$ $(k \neq j)$ and so define a basis function $\varphi_j(x,y)$ such that

$$\varphi_j(x_k, y_k) = \begin{cases} 1 & (j = k) \\ 0 & (j \neq k) \end{cases} \quad (1 \leq j, k \leq 4) \tag{4.21}$$

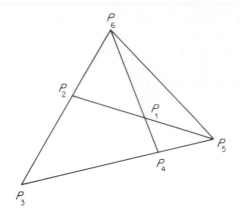

Figure 13

Exercise 13   Let P_5 be the intersection of $P_1 P_2$ and $P_3 P_4$ and let P_6 be the intersection of $P_2 P_3$ and $P_4 P_1$ (Figure 13). Verify that the line $P_5 P_6$ ($D_{56} = 0$) is such that

(i) $\quad \dfrac{D_{23}}{C_{123}} + \dfrac{D_{34}}{C_{134}} - \dfrac{D_{24}}{C_{124}} = \dfrac{D_{56}}{C_{156}}$ (4.22a)

and that there exists some constant $\lambda \neq 0$ for which

(ii) $\quad C_{134}C_{234}D_{12} + C_{123}C_{124}D_{34} =$

$\qquad C_{123}C_{234}D_{14} + C_{134}C_{124}D_{23} = \lambda D_{56}.$ (4.22b)

Hence show that if $\alpha C_{klm}C_{jkm} = \beta C_{jlm}C_{jkl}$ then the basis functions defined in (4.21) can be written as

$$\frac{D_{kl}}{C_{jkl}}\frac{D_{lm}}{C_{jlm}}\left(\frac{D_{56}}{C_{j56}}\right)^{-1}$$

where $P_k P_l$ and $P_l P_m$ are the sides of the quadrilateral not containing the corner P_j, and where (j,k,l,m) is some permutation of $(1,2,3,4)$. (Hint. Use the formulae

(i) $\quad C_{jkl} - C_{klm} + C_{jlm} - C_{jkm} = 0$ (4.23a)

and

(ii) $\quad C_{jkl} - D_{kl} + D_{jl} - D_{jk} = 0,$ (4.23b)

where (j,k,l,m) is any permutation of $(1,2,3,4)$.)

Isoparametric coordinates

Bilinear approximation

The most common method of using quadrilateral elements is to introduce a point transformation of the quadrilateral onto the unit square and then use the so-called isoparametric approximation (Irons, 1966; Zienkiewicz, 1971). That is, the corner points $P_j = (x_j, y_j)$ ($j = 1,2,3,4$) of the quadrilateral are transformed into the four points $(1,1)$, $(0,1)$, $(0,0)$ and $(1,0)$ (in (p,q) space). The standard transformation is

$$t = pqt_1 + (1-p)qt_2 + (1-p)(1-q)t_3 + p(1-q)t_4 \quad (t = x,y),$$

(4.24)

which can be written as

$$t = \sum_{j=1}^{4} \varphi_j^{(1)}(p,q) t_j \quad (t = x,y).$$

(4.25)

An *isoparametric approximation* is obtained by defining the approximation in a form similar to (4.25), namely

$$U(p,q) = \sum_{j=1}^{4} \varphi_j^{(1)}(p,q) U_j.$$

(4.26)

Exercise 14   Verify that the inverse transformation of (4.25) can be written as

$$(C_{234} + C_{134})p^2 + (D_{34} + D_{12} - C_{123} - C_{234})p + D_{23} = 0$$

(4.27a)

and

$$(C_{234} + C_{123})q^2 + (D_{23} + D_{41} - C_{134} - C_{234})q + D_{34} = 0.$$

(4.27b)

Further, by using (4.23a) and (4.23b), show that the function p defined by (4.27a) is equivalent to a surface of the form (4.20) through the points $\mathscr{P}_j = (x_j, y_j, f_j)$ with $f_2 = f_3 = 0$ and $f_1 = f_4 = 1$, if $\alpha = \beta$, and that q defined by (4.27b) is equivalent to a surface with $f_3 = f_4 = 0$, $f_1 = f_2 = 1$ if $\alpha = \beta$.

Exercise 15   The particular choice $\alpha = \beta$ is not the only one possible, if a transformation from the quadrilateral $P_1 P_2 P_3 P_4$ to the unit square is required. Show that if $\mathscr{P}_j$ ($j = 1,2,3,4$) are chosen as in Exercise 14 and $\alpha C_{klm} C_{jkm} = \beta C_{jlm} C_{jkl}$, then the new coordinates p and q can be

84

defined by

$$p = \frac{D_{23}}{C_{123}} \frac{D_{34}}{C_{134}} \left(\frac{D_{56}}{C_{156}}\right)^{-1} + \frac{D_{23}}{C_{234}} \frac{D_{12}}{C_{124}} \left(\frac{D_{56}}{C_{456}}\right)^{-1}$$

and

$$q = \frac{D_{23}}{C_{123}} \frac{D_{34}}{C_{134}} \left(\frac{D_{56}}{C_{156}}\right)^{-1} - \frac{D_{34}}{C_{234}} \frac{D_{14}}{C_{124}} \left(\frac{D_{56}}{C_{256}}\right)^{-1}.$$

Exercise 16 Show that the Jacobian J of the transformation defined by (4.25) can be written as

$$J = (1-p)C_{123} + (1-q)C_{134} + (p+q+1)C_{124}$$

and verify that $J > 0$ for $0 \leqslant p,q \leqslant 1$.

If the bilinear polynomials used in the transformation (4.25) are replaced by polynomials of higher degree, it is possible to introduce additional points in the transformation and at the same time extend the isoparametric approximations to curvilinear quadrilaterals.

Biquadratic approximation

In addition to the four points $P_j = (x_j, y_j)$ ($j = 1,2,3,4$) which correspond to the corners of the unit square in (p,q) space, we now consider the points P_j ($j = 5, \ldots, 9$) which correspond to the side mid-points $(\frac{1}{2},1)$, $(0,\frac{1}{2})$, $(\frac{1}{2},0)$ and $(1,\frac{1}{2})$ and the centre $(\frac{1}{2},\frac{1}{2})$ respectively (Figure 14). The biquadratic transformation is defined by

$$t = \sum_{j=1}^{9} \varphi_j^{(2)}(p,q) t_j \quad (t = x,y), \tag{4.28}$$

where $\varphi_1^{(2)} = p(2p-1)q(2q-1)$, with $\varphi_2^{(2)}$, $\varphi_3^{(2)}$ and $\varphi_4^{(2)}$ defined similarly,

$$\varphi_5^{(2)} = 4(1-p)pq(2q-1),$$

with $\varphi_6^{(2)}$, $\varphi_7^{(2)}$ and $\varphi_8^{(2)}$ defined similarly and with

$$\varphi_9^{(2)} = 16p(1-p)q(1-q).$$

The isoparametric approximation is then defined by

$$U(p,q) = \sum_{j=1}^{9} \varphi_j^{(2)}(p,q) U_j. \tag{4.29}$$

Alternatively if the transformation is defined by (4.25), and the approximation is defined by (4.29), then this is an example of *subparametric approximation*.

The sides of the quadrilateral can be made straight by suitable

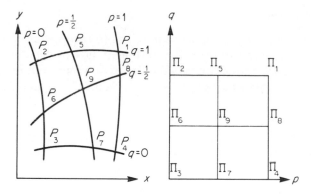

Figure 14

positioning of the side nodes P_5, P_6, P_7 and P_8. In particular if these are the mid-points of the respective sides and P_9 is taken as the centroid of the quadrilateral then formulae (4.28) reduce to formulae (4.25).

The internal node P_9 can be eliminated, in the same way as internal nodes were eliminated from approximations based on triangular elements, by using the linear relation

$$U_9 = \frac{1}{2} \sum_{j=5}^{8} U_j - \frac{1}{4} \sum_{j=1}^{4} U_j.$$

This yields a function that still interpolates quadratics *in p and q* exactly, but has *no term in* $p^2 q^2$ in the approximation which can be written as

$$U(p,q) = \sum_{j=1}^{8} \varphi_j^{(2)*}(p,q) U_j, \qquad (4.30)$$

where

$$\varphi_1^{(2)*} = pq(2p + 2q - 3)$$

and similarly for $\varphi_j^{(2)*}$ $(j = 2,3,4)$,

$$\varphi_5^{(2)*} = 4pq(1 - p)$$

and similarly for $\varphi_j^{(2)*}$ $(j = 6,7,8)$. For an eight-node isoparametric approximation, the transformation from (x,y) to (p,q) will also be defined by (4.30), with U replaced by x and y in turn. This eight-node quadrilateral has been used by Jordan (1970) as an element for use in solving problems involving plane stress or strain.

Bicubic approximation

The full bicubic approximation involves four internal nodes (Figure 15), in addition to the four corner and eight side nodes (two on each side). The internal nodes can be eliminated to yield an approximation, with no terms involving $p^2 q^2$, $p^3 q^2$, $p^2 q^3$ and $p^3 q^3$, that can be written as

$$U(p,q) = \sum_{j=1}^{12} \varphi_j^{(3)*}(p,q)U_j,$$

where

$$\varphi_1^{(3)*} = \tfrac{9}{2}(p^2 + q^2 - p - q + \tfrac{2}{9})pq$$

and similarly for $\varphi_j^{(3)*}$ $(j = 2,3,4)$,

$$\varphi_5^{(3)*} = \tfrac{9}{2}pq(1 - p)(3p - 1)$$

and similarly for $\varphi_j^{(3)*}$ $(j = 6, \ldots, 12)$.

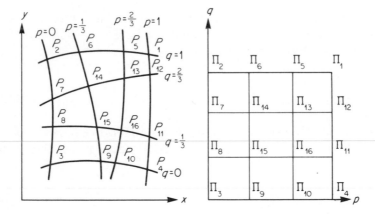

Figure 15

4.4 THE TETRAHEDRON (cf. The triangle)

Lagrange interpolation

The complete mth-order polynomial

$$\Pi_m(x,y,z) = \sum_{j+k+l=0}^{m} \alpha_{jkl}x^j y^k z^l,$$

can be used to interpolate a function $U(x,y,z)$ at $\tfrac{1}{6}(m + 1)(m + 2)(m + 3)$ symmetrically placed nodes in a tetrahedron. The first three cases of this general representation for the tetrahedron $P_1P_2P_3P_4$ are:

(1) *The linear case* ($m = 1$). The polynomial is

$$\Pi_1(x,y,z) = \sum_{j=1}^{4} U_j p_j^{(1)}(x,y,z),$$

where U_j ($j = 1,2,3,4$) are the values of $U(x,y,z)$ at the vertices P_j and

$$p_j^{(1)}(x,y,z) = \frac{1}{\Gamma_{jkln}}(E_{kln} - A_{kln}x + B_{kln}y - C_{kln}z)$$

where

$$\Gamma_{jkln} = \det \begin{bmatrix} 1 & x_j & y_j & z_j \\ 1 & x_k & y_k & z_k \\ 1 & x_l & y_l & z_l \\ 1 & x_n & y_n & z_n \end{bmatrix},$$

$$A_{kln} = \det \begin{bmatrix} 1 & y_k & z_k \\ 1 & y_l & z_l \\ 1 & y_n & z_n \end{bmatrix}$$

and so on, where (j,k,l,n) is any permutation of $(1,2,3,4)$. Note that the modulus of Γ_{jkln} is six times the volume of the tetrahedron $P_1P_2P_3P_4$.

(2) *The quadratic case* ($m = 2$). The polynomial is now

$$\Pi_2(x,y,z) = \sum_{j=1}^{10} U_j p_j^{(2)}(x,y,z),$$

where U_j ($j = 1,2,3,4$) are the values of $U(x,y,z)$ at the vertices P_j and U_j ($j = 5, \ldots, 10$) are the values at the mid-points of the edges. The formulae for $p_j^{(2)}$ in terms of $p_j^{(1)}$ have the same forms as for the triangle (Section 4.1).

(3) *The cubic case* ($m = 3$). The polynomial is now

$$\Pi_3(x,y,z) = \sum_{j=1}^{20} U_j p_j^{(3)}(x,y,z)$$

where U_j ($j = 1,2,3,4$) are as above, U_j ($j = 5, \ldots, 15$) are the values at the points of trisection of the sides and U_j ($j = 16, \ldots, 20$) are the values at the centroids of the faces. The formulae for $p_j^{(3)}$ in terms of $p_j^{(1)}$ have the same forms as for the triangle (Section 4.1).

Hermite interpolation

A cubic approximation to $U(x,y,z)$ in the tetrahedron can be written down in a manner analogous to (4.9)—(4.10b) as

$$G_1(x,y,z) = \sum_{j=1}^{4} \left\{ U_j q_j^{(3)} + \left(\frac{\partial U}{\partial x}\right)_j r_j^{(3)} + \left(\frac{\partial U}{\partial y}\right)_j s_j^{(3)} \right.$$

$$\left. + \left(\frac{\partial U}{\partial z}\right)_j t_j^{(3)} + \bar{U}_j \bar{q}_j^{(3)} \right\}, \tag{4.31}$$

where $\bar{U}_j$ ($j = 1, \ldots, 4$) are the values of $U(x,y,z)$ at the centroids of the faces opposite P_j and where

$$q_j^{(3)}(x,y,z) = p_j(3p_j - 2p_j^2 - 7(p_k p_l + p_l p_n + p_n p_k)), \tag{4.32a}$$
$$\bar{q}_j^{(3)}(x,y,z) = 27 p_k p_l p_n$$

and

$$r_j^{(3)}(x,y,z) = p_j[(\xi_{jk} p_k(p_l + p_n - p_j) + \xi_{jl} p_l(p_k + p_n - p_j)$$

$$+ \xi_{jn} p_n(p_k + p_l - p_j)]; \tag{4.32b}$$

$s_j^{(3)}(x,y,z)$ is obtained from $r_j^{(3)}(x,y,z)$ by replacing ξ by η and $t^{(3)}(x,y,z)$ by replacing ξ by ζ, where $\zeta_{jk} = z_j - z_k$. As for the triangle we have replaced $p_j^{(1)}$ by p_j ($j = 1,2,3,4$) to simplify the formulae.

Exercise 17   Verify that it is possible to eliminate the values of $U(x,y,z)$ at the centroids of the faces and still interpolate quadratic polynomials exactly, if the replacement is

$$\bar{U}_j = \frac{1}{3} \sum_{k \neq j} U_k + \frac{1}{18} \Sigma^{(1)} \left\{ \left(\frac{\partial U}{\partial x}\right)_k (\xi_{nk} + \xi_{lk}) + \left(\frac{\partial U}{\partial y}\right)_k (\eta_{nk} + \eta_{lk}) \right.$$

$$\left. + \left(\frac{\partial U}{\partial z}\right)_k (\zeta_{nk} + \zeta_{lk}) \right\},$$

where $\Sigma^{(1)}$ denotes the summation (given j) over (j,k,l,n) for all possible even permutations of $(1,2,3,4)$.

Show that the interpolating polynomial becomes

$$G_1^*(x,y,z) = \sum_{j=1}^{4} \left\{ U_j q_j^{(3)*} + \left(\frac{\partial U}{\partial x}\right)_j r_j^{(3)*} + \left(\frac{\partial U}{\partial y}\right)_j s_j^{(3)*} \right.$$

$$\left. + \left(\frac{\partial U}{\partial z}\right)_j t_j^{(3)*} \right\},$$

where

$$q_j^{(3)^*}(x,y,z) = p_j(3p_j - 2p_j^2 + 2(p_k p_l + p_k p_n + p_l p_n)),$$

$$r_j^{(3)^*}(x,y,z) = p_j^2(\xi_{kj} p_k + \xi_{lj} p_l + \xi_{nj} p_n) + \tfrac{1}{2} p_j \{ p_k p_l(\xi_{kj} + \xi_{lk})$$
$$+ p_k p_n(\xi_{kj} + \xi_{nj}) + p_l p_n(\xi_{lj} + \xi_{nj})\}$$

and so on.

4.5 THE HEXAHEDRON (cf. The quadrilateral)

The element having six quadrilateral faces is in general a better element in three dimensions than the tetrahedron. Local isoparametric coordinates (p,q,r) can be introduced as for quadrilateral elements in two dimensions. The transformations are given by

$$t = \sum_{j=1}^{8} \varphi_j^{(1)}(p,q,r) t_j \quad (t = x,y,z), \tag{4.33}$$

where

$$\varphi_1^{(1)} = pqr,$$
$$\varphi_2^{(1)} = (1-p)qr,$$

etc., where the vertices are numbered as in Figure 16. An arbitrary hexahedron is thus transformed into the unit cube in (p,q,r) space. The isoparametric approximation is then defined by

$$U(p,q,r) = \sum_{j=1}^{8} \varphi_j^{(1)}(p,q,r) U_j. \tag{4.34}$$

Triquadratic and tricubic approximations can be generated if

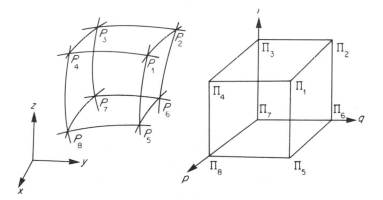

Figure 16

additional points on the sides and the faces are introduced, together
with a number of interior points. The face points and interior points
can be eliminated in a manner analogous to that used in two dimensions
to eliminate the interior points of a quadrilateral.

Exercise 18 Compute the basis functions $\varphi_j^{(2)}(p,q,r)$ $(j = 1, \ldots, 27)$
for the triquadratic isoparametric approximation defined for a
hexahedron. Then verify that the centre face points and the centroid
can be eliminated to yield an approximation of the form

$$U(p,q,r) = \sum_{j=1}^{20} \varphi_j^{(2)*}(p,q,r)U_j, \tag{4.35}$$

which interpolates quadratics exactly, has no terms in $p^2 q^2$, $q^2 r^2$, $r^2 p^2$,
$p^2 q^2 r$, $p^2 q r^2$, $p q^2 r^2$, $p^2 q^2 r^2$ and for which

$$\varphi_1^{(2)*} = pqr(2p + 2q + 2r - 5)$$

and similarly for $\varphi_j^{(2)*}$ $(j = 2, \ldots, 8)$,

$$\varphi_9^{(2)*} = 4pqr(1 - p),$$

with $\varphi_j^{(2)*}$ $(j = 10, \ldots, 20)$ similarly.

Exercise 19 Verify that tricubic approximation in a hexahedron
requires 64 nodes, one at each corner with a further two on each edge,
four on each face and eight in the interior of the hexahedron. Then
verify that the face points and interior points can be eliminated to yield
an approximation that involves terms in $p^j q^k r^l$ $(j + k + l \leqslant 4)$ together
with the three quintic terms $p^3 qr$, $pq^3 r$ and pqr^3, such that

$$U(p,q,r) = \sum_{j=1}^{32} \varphi_j^{(3)*}(p,q,r)U_j, \tag{4.36}$$

where

$$\varphi_1^{(3)*} = \tfrac{9}{2}pqr[p^2 + q^2 + r^2 - (p + q + r) + \tfrac{2}{9}],$$

with $\varphi_j^{(3)*}$ $(j = 2, \ldots, 8)$ similarly, and

$$\varphi_9^{(3)*} = \tfrac{9}{2}pqr(1 - p)(3p - 1),$$

with $\varphi_j^{(3)*}$ $(j = 10, \ldots, 32)$ similarly.

4.6 CURVED BOUNDARIES

So far basis functions have been constructed in the main for networks with straight sides. In real problems in two and three dimensions,
however, boundaries and interfaces are often curved. It is the purpose
of this section to derive basis functions for networks composed of
elements with curved sides (two dimensions) or curved surfaces (three
dimensions). The curved element was introduced into structural

analysis by Ergatoudis, Irons and Zienkiewicz (1968) and reference to it can be found in Zienkiewicz (1971). In two dimensions, elements with straight sides, usually triangles or quadrilaterals, are perfectly satisfactory if the domain has a polynomial boundary. If some part of the boundary is curved, however, elements with at least one curved side are desirable. This is also the case when curved material interfaces are present in the region.

Initially we consider the triangular element with two straight sides and one curved side. This element together with triangles with straight sides can deal adequately with most plane problems involving curved boundaries and interfaces.

(A) Triangles with one curved side

The triangle $P_1 P_2 P_3$ is considered in the (x,y) plane with $l(x,y) - 0$ and $m(x,y) = 0$ the equations of the straight sides $P_2 P_3$ and $P_3 P_1$ respectively. The equation of the curved side passing through P_1 and P_2 is $F(x,y) = 0$. We take $l(x,y)$, $m(x,y)$ and $F(x,y)$ to be normalized so that

$$l(x_1,y_1) = m(x_2,y_2) = F(x_3,y_3) = 1.$$

The transformation from the (x,y) plane to the (l,m) is given by

$$l = \frac{1}{C_{123}} (\tau_{23} + \eta_{23}x - \xi_{23}y) \qquad (4.37a)$$

and

$$m = \frac{1}{C_{123}} (\tau_{31} + \eta_{31}x - \xi_{31}y). \qquad (4.37b)$$

In the (l,m) plane, the triangle becomes $P_1' P_2' P_3'$ where $P_1' = (1,0)$, $P_2' = (0,1)$ and $P_3' = (0,0)$, and the curved side $P_1' P_2'$ is given by

$$F(x(l,m), y(l,m)) = f(l,m) = 0.$$

We are using l and m in place of $p_1^{(1)}$ and $p_2^{(1)}$ (cf. (4.2)), because triangle $P_1' P_2' P_3'$ with one curved side is not the standard triangle.

We shall now describe one type of Lagrangian approximation defined on the curved triangle $P_1' P_2' P_3'$ which satisfies the following conditions:

(i) Linear polynomials are interpolated exactly, that is,

$$\sum_i \varphi_i = 1; \qquad \sum_i l_i\varphi_i = l; \qquad \sum_i m_i\varphi_i = m, \qquad (4.38)$$

where $\varphi_i(l,m)$ is a basis function associated with the node i.
Since the transformation from (x,y) to (l,m) is linear (4.38)

implies that linear polynomials in (x,y) on $P_1 P_2 P_3$ are also interpolated exactly.

(ii) The basis function $\varphi_3(l,m)$ corresponding to P_3' is identically zero on the curved side $P_2' P_1'$.

(iii) The resulting piecewise smooth function, defined on a network of triangles, each triangle with at most one curved side, is C^0 continuous.

It is not difficult to see that in order to satisfy (i) and (ii) it is necessary to have at least four nodes in the triangle $P_1' P_2' P_3'$. In the simplest case these are taken as the three vertices together with an additional point $P_4' = (l_4, m_4)$, on the curved side.

We first construct a basis function φ_3 that satisfies (ii) and then use (4.38) to construct φ_1, φ_2 and φ_4. We employ geometrical considerations similar to those adopted to obtain basis functions for the quadrilateral and so we consider the family of surfaces $z(l,m) = 0$ which intersect the (l,m) plane in the curve $f(l,m) = 0$ and are given by the equation

$$z(\alpha z + \beta l + \gamma m + \delta) + f(l,m) = 0. \tag{4.39}$$

If we impose conditions

(1) $z = 1$ $\qquad (l = m = 0)$,

(2) $z = 1 - l$ $\qquad (m = 0)$,

(3) $z = 1 - m$ $\qquad (l = 0)$,

then it is possible to specify β, γ and δ in terms of an arbitrary constant α, such that (4.39) becomes

$$\alpha z^2 + \left[\alpha(l + m - 1) + 1 - \frac{f(l,0)}{1 - l} - \frac{f(0,m)}{1 - m} \right] z + f(l,m) = 0. \tag{4.40}$$

We now identify z with φ_3 and the remaining basis functions are determined from (4.38) as

$$\varphi_1 = \frac{(1 - m_4)l + l_4 m - l_4}{1 - l_4 - m_4} + \frac{l_4}{1 - l_4 - m_4} \varphi_3,$$

$$\varphi_2 = \frac{m_4 l + (1 - l_4)m - m_4}{1 - l_4 - m_4} + \frac{m_4}{1 - l_4 - m_4} \varphi_3 \tag{4.41}$$

and

$$\varphi_4 = \frac{1 - l - m}{1 - l_4 - m_4} - \frac{1}{1 - l_4 - m_4} \varphi_3.$$

Exercise 20   Verify that if $\alpha = 0$ in (4.40) the basis function φ_3

becomes

$$\varphi_3 = \frac{f(l,m)}{f(0,m)/(1-m) + f(l,0)/(1-l) - 1},$$ (4.42)

and further that if the curved side is part of the conic

$$f(l,m) = al^2 + blm + cm^2 - (1+a)l - (1+c)m + 1 = 0,$$

then

$$\varphi_3 = \frac{al^2 + blm + cm^2 - (1+a)l - (1+c)m + 1}{1 - al - cm}$$ (4.43)

and we recover the rational basis functions of Wachspress (1971, 1973, 1974 and 1975).

Exercise 21 If the curved side is part of a hyperbola

$$f(l,m) = blm - l - m + 1 = 0,$$

show that the basis functions φ_i ($i = 1,2,3,4$) are polynomials if $\alpha = 0$.

Note. Piecewise hyperbolic arcs can be used to approximate a curved interface or boundary and still permit polynomial basis functions.

(B) Isoparametric coordinates (cf. Section 4.3)

Quadratic approximation in the standard triangle

Triangular elements with curved boundaries are often dealt with by introducing a point transformation onto the standard triangle and then using isoparametric approximations. These will be illustrated first with respect to the general curved triangle shown in Figure 17. The point transformation from the (x,y) plane to the (p,q) plane is given by

$$t = \sum_{j=1}^{6} p_j^{(2)}(p,q)t_j \quad (t = x,y),$$ (4.44)

where the quadratic basis functions $p_j^{(2)}$ ($j = 1, \ldots, 6$) are as defined in Section 4.1 with p_1 and p_2 replaced by p and q respectively. An isoparametric approximation is then obtained by defining the approximation in a similar form, that is,

$$U(p,q) = \sum_{j=1}^{6} p_j^{(2)} U_j.$$ (4.45)

If we consider the special case of a triangle with two straight sides and one curved side such that $t_5 = \frac{1}{2}(t_2 + t_3)$, $t_6 = \frac{1}{2}(t_3 + t_1)$ ($t = x,y$), the transformation (4.44) reduces to

$$l = \alpha pq + p$$ (4.46a)

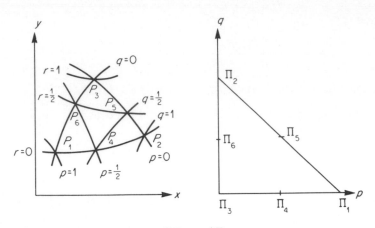

Figure 17

and

$$m = \beta pq + q, \tag{4.46b}$$

where $\alpha = 2(2l_4 - 1)$ and $\beta = 2(2m_4 - 1)$. The inverse transformation is given by

$$\beta p + (\beta l - \alpha m + 1)q - 1 = 0 \tag{4.47a}$$

and

$$\alpha q + (\alpha m - \beta l + 1)p - m = 0. \tag{4.47b}$$

Exercise 22   Show that if $l_4 = m_4 = R$ then it follows from (4.47a) and (4.47b) that $r (= 1 - p - q)$ is given by

$$r^2 - \frac{4R-1}{2R-1}r + \left[\frac{2R-l-m}{2R-1} - (l-m)^2\right] = 0, \tag{4.47c}$$

and hence that the curved side $f(l,m) = 0$ is replaced by the curve $r = 0$ given by

$$1 - \frac{l+m}{2R} - \frac{2R-1}{2R}(l-m)^2 = 0, \tag{4.48}$$

which is a parabola. Further show that if $\alpha = (2R - 1)/2R$ and $f(l,m)$ is given by (4.48) then (4.40) becomes (4.47c) with r replacing z. Explain this link between isoparametric approximation and direct methods of dealing with curved boundaries.

Exercise 23   Show that in general the curved side $r = 0$ is given by

$$(\beta l - \alpha m)^2 + (\alpha + \beta + \alpha\beta - \beta^2)l + (\alpha + \beta + \alpha\beta - \alpha^2)m$$
$$- (\alpha + \beta + \alpha\beta) = 0,$$

where α and β are as given above.

Forbidden elements

One of the difficulties in dealing with curved elements using isoparametric coordinates arises from the vanishing of the Jacobian $J (= 1 + \beta p + \alpha q)$ of the transformation defined by (4.46a) and (4.46b). This Jacobian is positive for all p,q such that $0 \leqslant p,q$ and $p + q \leqslant 1$ provided that the point (l_4, m_4) lies in the region $l,m > \frac{1}{4}$ as shown in Figure 18. For other positions of the point (l_4, m_4) in the positive quadrant of the (l,m) plane, including the lines $l = \frac{1}{4}$, $m = \frac{1}{4}$, the Jacobian either vanishes or is negative for certain values of (p,q) (Jordan, 1970) and so isoparametric coordinates cannot in general be used to deal with curved elements in these 'forbidden' cases, exceptions to this rule are given in Section 7.4(F). This is because results calculated in terms of the isoparametric coordinates (p,q) cannot be transferred back to the (l,m) plane because of the vanishing of the Jacobian of the transformation somewhere in the element.

Instead of using isoparametric coordinates, we can deal directly in terms of l and m by using (4.40) and (4.41). As in Exercise 22, if the curved side is given by (4.48) and $\alpha = (2R - 1)/2R$, then (4.40) will have real roots provided

$$F(l,m;R) = \left(1 + m + \frac{1}{2(2R - 1)}\right)^2 - 4lm \geqslant 0. \tag{4.49}$$

It is easy to see that for a fixed value of R the function $F(l,m;R)$ has no maximum or minimum inside the element, or indeed anywhere in the (l,m) plane. Consequently the smallest value of $F(l,m;R)$ will occur on the boundary of the element for all values of R. In fact, using (4.48), it follows that

$$F(l,m;R) = \left(\frac{4R - 1}{4R - 2}\right)^2 > 0$$

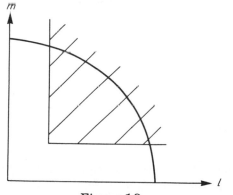

Figure 18

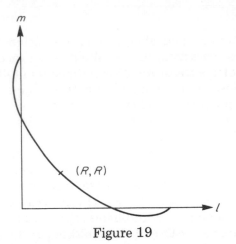

Figure 19

for all R on the curved side, and of course F is always positive from (4.49) on $l = 0$ and $m = 0$. The condition (4.49) is thus satisfied for all values of R and for all points (l,m) in the element.

When isoparametric coordinates are used in this example, values of R in the range $0 < R < \frac{1}{4}$ are forbidden. This does not seem to be the case when (4.40) is used. However, by simple geometrical considerations, it can be shown that for $0 < R < \frac{1}{4}$, the curved boundary intersects the l and m axes at points between the origin and the unit points (Figure 19).

Exercise 24 For $0 < R < \frac{1}{4}$, find the points on the l and m axes between the origin and the unit points where the curved side intersects the axes, and show that as $R \to \frac{1}{4}$ these points tend towards the unit points on the respective axes.

Exercise 25 Show that the quadratic equation (4.40) has real roots for all values of α when (l,m) is a point on the perimeter of the triangle with two straight sides and one curved side.

Cubic approximation in the standard triangle

The unique cubic interpolating polynomial in the (p,q) plane can be written as

$$U(p,q) = \sum_{j=1}^{10} p_j^{(3)} U_j, \tag{4.50}$$

where the basis functions $p_j^{(3)}$ $(j = 1, \ldots, 10)$ are given in Section 4.1. An isoparametric approximation is obtained if (4.50) is used to define the point transformation from (x,y) (or (l,m)) to (p,q) by replacing U

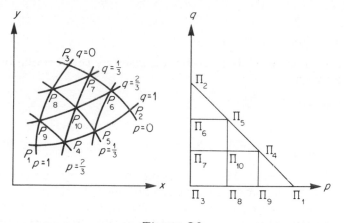

Figure 20

by x and y (or l and m) (Figure 20). For the special case of a triangle with two straight sides such that P_6, P_7 and P_8, P_9 are the points of trisection of $P_2 P_3$ and $P_3 P_1$ respectively, these formulae reduce to

$$l = p + \tfrac{9}{2} pq(6l_{10} - l_4 - l_5 - 1) + 2\tfrac{7}{2} p^2 q(l_4 - 2l_{10})$$
$$+ 2\tfrac{7}{2} pq^2 (l_5 - 2l_{10} + \tfrac{1}{3})$$

and   (4.51)

$$m = q + \tfrac{9}{2} pq(6m_{10} - m_4 - m_5 - 1) + 2\tfrac{7}{2} p^2 q(m_4 - 2m_{10} + \tfrac{1}{3})$$
$$+ 2\tfrac{7}{2} pq^2 (m_5 - 2m_{10}),$$

where $P_j' = (l_j, m_j)$ $(j = 4, 5)$ are points on the curved side and (l_{10}, m_{10}) is inside the triangle. From (4.51), it follows that the replacement curve passing through the points P_1', P_4', P_5' and P_2' is a *cubic* curve.

Exercise 26   Show that if

$$l_4 = l_5 + \tfrac{1}{3}$$

and

$$m_4 = m_5 - \tfrac{1}{3},$$

then the cubic curve degenerates into a *unique* parabola through the four points $(1,0)$, (l_4, m_4), (l_5, m_5) and $(0,1)$. If in addition

$$l_4 = 2l_{10}$$

and

$$m_5 = 2m_{10},$$

prove that the transformation formulae (4.51) reduce to

$$l = p + 9(l_{10} - \tfrac{1}{3})pq$$

98

and

$$m = q + 9(m_{10} - \tfrac{1}{3})pq,$$

and that the equation of the parabola is given by the formula in Exercise 23 with

$$4l_4 = 9l_{10} - 1$$

and

$$4m_4 = 9m_{10} - 1.$$

In McLeod and Mitchell (1975), examples of arbitrary curved sides are chosen and matching parabolic arcs are obtained. In all examples a parabola close to the original curve was found. It is worth pointing out that isoparametric elements in general are extremely sensitive to distortion from the basic triangular shape.

(C) Quadrilateral with one curved side

In Section 4.3 the formula for biquadratic approximation on a quadrilateral was given by (4.30). For the special case of a quadrilateral with three straight sides and one curved side (Figure 21) where P_6, P_7 and P_8 are the mid-points of the straight sides P_2P_3, P_3P_4 and P_4P_1 respectively, the point transformation formulae corresponding to (4.30) reduce to

$$l = p + (-1 - l_1 + 4l_5)pq + (2l_1 - 4l_5)p^2 q$$

and $\qquad\qquad\qquad\qquad\qquad\qquad\qquad\qquad\qquad$ (4.52)

$$m = q + (-3 - m_1 + 4m_5)pq + (2 + 2m_1 - 4m_5)p^2q,$$

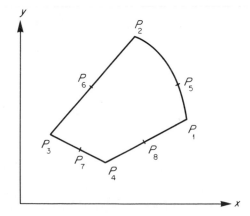

Figure 21

for an isoparametric approximation, where

$$l = \frac{1}{C_{234}}(\tau_{23} + \eta_{23}x - \xi_{23}y)$$

and

$$m = \frac{1}{C_{234}}(\tau_{34} + \eta_{34}x - \xi_{34}y).$$

From (4.52), the coordinate q can be eliminated to give

$$Tp^3 + [Z - Tl + Ym]p^2 + [1 + Xm - Zl]p - l = 0.$$

The curve $q = 1$ is given by

$$[Tl + Y(1 - m)]^2 + (T + XT - YZ)[Zl + (1 + X)(1 - m)] = 0, \tag{4.53}$$

where

$$X = -1 - l_1 + 4l_5,$$
$$Y = 2l_1 - 4l_5,$$
$$Z = -3 - m_1 + 4m_5$$

and

$$T = 2 + 2m_1 - 4m_5.$$

This curve is of course a *parabola*. Hence if isoparametric coordinates are used, as defined by the point transformation formulae corresponding to (4.30), *the curved side is replaced by the parabolic arc, whose equation is given by (4.53)*.

Exercise 27 For the case of a quadrilateral with one curved side and two equally spaced intermediate nodes on each straight side (Figure 22) show that the point transformation formulae reduce to

$$l = p + \tfrac{9}{2}[(\tfrac{2}{9}l_1 - l_5 + 2l_6 - \tfrac{2}{9})pq - (l_1 - 4l_5 + 5l_6)p^2 q$$
$$+ (l_1 - 3l_5 + 3l_6)p^3 q]$$

and

$$m = q + \tfrac{9}{2}[(\tfrac{2}{9}m_1 - m_5 + 2m_6 - \tfrac{11}{9})pq - (m_1 - 4m_5 + 5m_6 - 2)p^2 q$$
$$+ (m_1 - 3m_5 + 3m_6 - 1)p^3 q].$$

Then show that the isoparametric coordinate q can be eliminated to give a quartic in p, the curve $q = 1$ is a quartic in l and m, and the curves of constant p are straight lines.

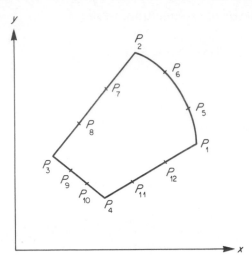

Figure 22

(D) Tetrahedron with curved faces

Consider a tetrahedron with four curved faces and one intermediate node on each of the six curved sides. This is illustrated in Figure 23(a). The standard tetrahedron in (p,q,r) space can be transformed into an arbitrary curved tetrahedron in (x,y,z) space with vertices and intermediate nodes in common with the original curved tetrahedron by using the point transformation formulae

$$t = p(2p-1)t_1 + q(2q-1)t_2 + r(2r-1)t_3 + s(2s-1)t_4 + 4pst_5$$
$$+ 4pqt_8 + 4qst_6 + 4rqt_9 + 4prt_{10} + 4rst_7 \quad (t = x,y,z), \quad (4.54)$$

where $p + q + r + s = 1$. For an isoparametric element, the interpolating function is also given by (4.54) with U replacing t. Since most finite regions in three-dimensional space can be divided up approximately† into tetrahedral elements either with four plane faces or with one curved and three plane faces, we shall concentrate on the latter tetrahedron. If the plane faces $P_2'P_3'P_4'$, $P_3'P_1'P_4'$ and $P_1'P_2'P_4'$ are represented by $l = 0$, $m = 0$ and $n = 0$ respectively with the vertices P_1', P_2' and P_3' taking the values $l = 1$, $m = 1$ and $n = 1$ respectively, then we arrive at Figure 23(b), where P_5', P_6' and P_7' are mid-points of the respective straight sides, and $P_8' = (R_1,R_1,0)$, $P_9' = (0,R_2,R_2)$ and $P_{10}' = (R_3,0,R_3)$.

---

†Such a division duplicates pieces on the surface of the region which are shaped like 'orange segments'.

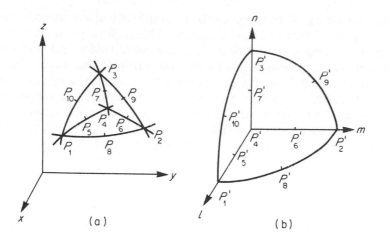

Figure 23

From (4.54) the point transformation formulae reduce to

$$l = p + 2(2R_1 - 1)pq + 2(2R_3 - 1)pr,$$
$$m = q + 2(2R_1 - 1)pq + 2(2R_2 - 1)qr \qquad (4.55)$$

and

$$n = r + 2(2R_3 - 1)pr + 2(2R_2 - 1)qr.$$

After considerable manipulation, it can be shown that *the surface* $1 - p - q - r = 0$ *is a quartic in l, m and n.*

(E) Hexahedron with curved faces

A finite region in three dimensions enclosed by a curved surface can be divided up into a finite number of hexahedral elements with curved faces. Point transformations based on isoparametric approximations such as (4.35) and (4.36) can be used to transform an arbitrary hexahedron with curved faces into the unit cube in (p,q,r) space. Calculations then proceed in the transformed space in the usual manner. Unfortunately the point transformation formulae corresponding to (4.35) and (4.36) are so complicated that it is impossible to obtain from them any indication of the shape of the curved surface implied by the point transformations. The only comforting thought is that it can be made to pass through a large number of points which lie on the original curved surface.

Note. There is a great need for basis functions which can cope *exactly* with specific curved sides and surfaces in two and three dimensions respectively. This is particularly true in inviscid fluid dynamics where 'tampering' with the boundary completely alters the velocity pattern in the vicinity of the boundary. Although isoparametric techniques are an improvement over replacing curved sides and surfaces by straight sides and planes respectively, they are still based on point transformations and so only give approximations to the original curves and surfaces. Preliminary work based on geometrical considerations for finding exact basis functions for curved elements can be found in Wachspress (1975), McLeod (1977), McLeod and Mitchell (1972) and Barnhill and Gregory (1976a) and (1976b).

Chapter 5

Convergence of Approximation

5.1 INTRODUCTION

In Chapter 3 the Galerkin approximation was introduced. It was shown how such an approximation U satisfies an equation which can be written as

$$a(U,V) = (f,V) \quad \text{(for all } V \in K_N\text{)}, \tag{5.1}$$

where K_N is an N-dimensional subspace of the space of admissible functions $\mathcal{H}$. If it is assumed that the integrals are evaluated exactly, an analysis of the errors for piecewise smooth approximations of the form

$$U(\mathbf{x}) = \sum_{i=1}^{N} \alpha_i \varphi_i(\mathbf{x})$$

reduces to two distinct steps:

(1) A proof of the best or near-best properties of the approximation. In Section 3.5, it was shown that the Ritz approximation is best in the energy norm; that is,

$$\| u - U \|_A = \inf_{\tilde{u} \in K_N} \| u - \tilde{u} \|_A. \tag{5.2}$$

In general a more useful result is that a Galerkin approximation given by (5.1) is *near best* in some *Sobolev norm*;[†] that is if

$$\| u - U \|_{r,R} \leqslant C \inf_{\tilde{u} \in K_N} \| u - \tilde{u} \|_{r,R} \tag{5.3}$$

for some $r, C > 0$.[‡]

(2) Prescribing an upper bound on the right-hand side of (5.3) by considering the particular case when $\tilde{u} \in K_N$ interpolates the solution. When K_N is a space of finite element approximations, there is usually an integer $k = k(K_N)$ and a constant $C = C(K_N)$

---

†Sobolev spaces and Sobolev norms are defined on page 105.
‡In this chapter, C is used to denote a positive constant which will not be the same at each occurrence.

such that the error in interpolation is bounded as

$$\| u - \tilde{u} \|_{r,R} \leqslant Ch^{k+1-r} \| u \|_{k+1,R} \quad (r = 0,1,\ldots,k) \quad (5.4)$$

provided $u \in \mathscr{H}_2^{(k+1)}(R)$. If it is assumed that the solution u satisfies

$$a(u,v) = (f,v) \quad (\text{for all } v \in \mathscr{H})$$

and is an element of the Sobolev space $\mathscr{H}_2^{(k+1)}(R)$, inequalities (5.3) and (5.4) can be combined to provide an estimate of the order of convergence of the Galerkin approximation U, as h tends to zero.

If, as usually happens, the integrals are evaluated numerically using a quadrature rule, then the approximate solution is no longer derived from (5.1) but from perturbed equations. The modified form can be written as

$$a_h(U_h,V) = (f,V)_h \quad (\text{for all } V \in K_N). \quad (5.5)$$

In such circumstances it is possible to derive an estimate of the order of convergence using (5.3) and (5.4) if a bound is available in the form

$$\| U - U_h \|_{r,R} \leqslant Ch^s$$

for some $s > 0$. Perturbations of the boundary also lead to modified equations in the form (5.5), as does the use of non-conforming elements, i.e. inadmissible functions such that $U_h \notin \mathscr{H}$. So it is clear that an analysis of problems defined by (5.5) is of considerable importance in the study of *practical* finite element methods.

Notation and preliminaries

A bilinear form a, is said to be $\mathscr{H}$-*elliptic*† if there exists a constant $\gamma > 0$ such that for all $u \in \mathscr{H}$

$$a(u,u) \geqslant \gamma \| u \|^2.$$

Following Chapter 1, it is said to be *bounded* if there exists a constant $\alpha > 0$, such that for all $u,v \in \mathscr{H}$,

$$| a(u,v) | \leqslant \alpha \| u \| \| v \|.$$

The majority of error estimates in this chapter are expressed in terms of *Sobolev norms*. Using the multi-index notation of the earlier chapters, these norms are written in the form

$$\| u \|_{k,R}^2 = \sum_{|i| \leqslant k} \| D^i u \|_{\mathscr{L}_2(R)}^2.$$

†Sometimes called *coercive* or *positive (definite)*.

Use is also made of semi-norms in terms of derivatives of one particular order; that is

$$|u|^2_{k,R} = \sum_{|i|=k} \|D^i u\|^2_{\mathscr{L}_2(R)}.$$

The Sobolev space $\mathscr{H}_2^{(k)}(R)$ is the space of all functions for which the corresponding Sobolev norm is finite. Unless otherwise stated (u,v) denotes the $\mathscr{L}_2(R)$ inner product of the functions $u(\mathbf{x})$ and $v(\mathbf{x})$; that is,

$$(u,v) = \iint_R u(\mathbf{x})v(\mathbf{x})d\mathbf{x}.$$

The *dual* of $\mathscr{H}_2^{(k)}(R)$ is denoted by $\mathscr{H}_2^{(-k)}(R)$ and the corresponding norm by

$$\|F\|_{-k,R} = \sup_{u \in \mathscr{H}_2^{(k)}(R)} \left\{ \frac{|F(u)|}{\|u\|_{k,R}} \right\}.$$

In Section 5.4(B) when $F \in \mathscr{H}_2^{(-k)}(R)$ is defined for some $v \in \mathscr{H}_2^{(k)}(R)$ as

$$F(u) = (u,v) \quad \text{(for all } u \in \mathscr{H}_2^{(k)}(R)\text{)},$$

it is possible, without ambiguity, to define

$$\|v\|_{-k,R} = \sup_{u \in \mathscr{H}_2^{(k)}(R)} \left\{ \frac{|(u,v)|}{\|u\|_{k,R}} \right\}.$$

Exercise 1 Prove that for any $u \in \mathscr{H}_2^{(k)}(R)$

$$\|u\|_{-k,R} \leqslant \|u\|_{\mathscr{L}_2(R)}.$$

In the following inequalities, the constants denoted by C may depend on the *region R* and the *space $\mathscr{H}$* of admissible functions, but not on the particular function involved. It is assumed that the region R is an open bounded domain, with a (piecewise) smooth boundary ∂R, that $\bar{R} = R \cup \partial R$ and that, unless otherwise stated, $R \subset \mathbb{R}^2$. For sufficiently smooth functions — say in $C^k(\bar{R})$ — use will occasionally be made of a *maximum norm* of the form

$$\|u\|_{(k)\bar{R}} = \max_{|i|\leqslant k}\{\|D^i u\|_{\mathscr{L}_\infty(\bar{R})}\},$$

together with the corresponding semi-norm

$$|u|_{(k)\bar{R}} = \max_{|i|=k}\{\|D^i u\|_{\mathscr{L}_\infty(\bar{R})}\}.$$

The space $\overset{\circ}{\mathscr{H}}_2^{(k)}(R)$ is defined for any $k > 0$ by

$$\overset{\circ}{\mathscr{H}}_2^{(k)}(R) = \left\{ u : u \in \mathscr{H}_2^{(k)}(R),\ u = \frac{\partial u}{\partial n} = \cdots = \frac{\partial^{k-1}u}{\partial n^{k-1}} = 0 \text{ on } \partial R \right\}$$

and similarly for any finite-dimensional subspace $K_N \subset \mathscr{H}_2^{(k)}(R)$ define

$$\overset{\circ}{K}_N = K_N \cap \mathscr{\overset{\circ}{H}}_2^{(k)}(R),$$

$\bar{K}_N$ denotes the complement of $\overset{\circ}{K}_N$ in K_N. For example for any 'finite element subspace' K_N, $\bar{K}_N$ is the subspace spanned by basis functions corresponding to the boundary nodes only; that is, basis functions which are zero at all the internal nodes.

In the interests of brevity, various additional restrictions on the region R will be omitted from the statements of the following lemmas. The restrictions are not mutually exclusive and are in general satisfied if the boundary is sufficiently smooth between the corners. The interested reader should consult the references cited for the details in each case. In particular it is assumed that *Sobolev's lemma* (for example Agmon, 1965, p. 32; or Yosida, 1965, p. 174) can be applied in $R \subset \mathbb{R}^m$ and hence that if $k > r + m/2$ then

$$\| u \|_{(r)\bar{R}} \leqslant C \| u \|_{k,R}.$$

Lemma 5.1

$$\| u \|_{\mathscr{L}_2(R)} \leqslant C \, | u \, |_{1,R} \quad (for \ all \ u \in \mathscr{\overset{\circ}{H}}_2^{(1)}(R)).$$

If $\lambda \, (>0)$ is the minimum eigenvalue of the Laplacian differential operator, subject to homogeneous Dirichlet boundary conditions, then $C = 1/\lambda$ (Courant and Hilbert, 1953, pp.398—404). It follows from Lemma 5.1 that on $\mathscr{\overset{\circ}{H}}_2^{(1)}(R)$, $\| \, . \, \|_{1,R}$ and $| \, . \, |_{1,R}$ are equivalent norms. Lemma 5.1 is a special case of more general results:

Lemma 5.2   (Aubin, 1972, p. 180)

$$\| u \|_{\mathscr{L}_2(R)} \leqslant C \, | u \, |_{k,R} \quad (for \ all \ u \in \mathscr{\overset{\circ}{H}}_2^{(k)}(R)).$$

Lemma 5.3   (Nečas, 1967, p. 18)

$$\| u \|_{k,R}^2 \leqslant C \left\{ | u \, |_{k,R}^2 + \sum_{|i| < k} \left| \iint D^i u \, dx \right|^2 \right\} \quad (for \ all \ u \in \mathscr{H}_2^{(k)}(R)).$$

Theorem 5.1   *The Bramble—Hilbert Lemma* (Bramble and Hilbert, 1970; Ciarlet and Raviart 1972a.) *Let $F \in \mathscr{H}_2^{(-k-1)}(R)$ be such that $F(p) = 0$ for all $p \in P_k$, then there exists some constant $C = C(R)$ such that*

$$| F(u) \, | \leqslant C \, \| F \|_{-k-1,R} \, | u \, |_{k+1,R}$$

for any $u \in \mathscr{H}_2^{(k+1)}(R)$.

Proof   It can be shown (Exercise 2), that for any $u \in \mathscr{H}_2^{(k+1)}(R)$ there exists a polynomial $p = p(u) \in P_k$ such that

$$\iint_R D^i(u + p) dx = 0 \quad (| \, i \, | \leqslant k).$$

Thus from Lemma 5.3

$$\| u + p \|_{k+1,R}^2 \leqslant C \mid u + p \mid_{k+1,R}^2 = C \mid u \mid_{k+1,R}^2 .$$

Then as the functional F is linear

$$F(u) = F(u + p)$$

and so combining all the results

$$\mid F(u) \mid \leqslant \| F \|_{-k-1,R} \| u + p \|_{k+1,R} \leqslant C \| F \|_{-k-1,R} \mid u \mid_{k+1,R} .$$

Exercise 2   Prove by induction on k, that for any $u \in \mathcal{H}_2^{(k+1)}(R)$ there exists a polynomial $p = p(u) \in P_k$ such that

$$\iint_R D^i(u + p)dx = 0.$$

The most frequent use of the Bramble–Hilbert lemma is in deriving bounds on bilinear forms. For example, let $\mathcal{H}$ be a Hilbert space and let F be a bounded bilinear form with arguments in $\mathcal{H}_2^{(k+1)}(R)$ and $\mathcal{H}$, that is, $F \in \mathcal{L}(\mathcal{H}_2^{(k+1)}(R) \times \mathcal{H}; \mathbb{R})$. Then if

$$F(u,v) = 0$$

for all $v \in \mathcal{H}$ when $u \in P_k$ and for some $v \in \mathcal{H}$, a functional $F_1 \in \mathcal{H}_2^{(-k-1)}(R)$ is defined by

$$F_1(u) = F(u,v) \quad \text{(for all } u \in \mathcal{H}_2^{(k+1)}(R)),$$

the Bramble–Hilbert lemma leads to

$$\mid F_1(u) \mid \leqslant C \| F_1 \|_{-k-1,R} \mid u \mid_{k+1,R} .$$

In Section 1.2 (Exercise 15), it is shown that for such a functional

$$\| F_1 \| \leqslant \| F \| \| v \|_{\mathcal{H}},$$

hence combining these results it follows that

$$\mid F(u,v) \mid \leqslant C \| F \| \| v \|_{\mathcal{H}} \mid u \mid_{k+1,R} . \tag{5.6}$$

Exercise 3   Using Lemma 5.3, prove that if $F \in \mathcal{L}(\mathcal{H}_2^{(k+1)}(R) \times \mathcal{H}_2^{(r+1)}(R); \mathbb{R})$ is such that

$$F(u,v) = 0 \quad \begin{cases} \text{for all } u \in \mathcal{H}_2^{(k+1)}(R) \text{ if } v \in P_r \\ \text{for all } v \in \mathcal{H}_2^{(r+1)}(R) \text{ if } u \in P_k, \end{cases}$$

then

$$\mid F(u,v) \mid \leqslant C \| F \| \mid u \mid_{k+1,R} \mid v \mid_{r+1,R} .$$

Regular transformations

It is usual to define basis functions on a standard element T_0, which could be either a unit square or a unit right triangle, a point transformation is then introduced to construct basis functions on an arbitrary element T (cf. Chapter 4). Thus it is natural to derive error estimates in terms of functions on a standard element, provided it is possible to transfer the bounds onto arbitrary elements.

In the case of curved elements, it is usual to consider the transformation in two parts by introducing an intermediate element T'. This intermediate element has the same vertices as the curved element T, but it has straight sides. Thus the mapping from T_0 onto T' is a linear transformation of the form

$$l = F_0(p)$$

such that

$$t = t_3 + (t_1 - t_3)p + (t_2 - t_3)q \quad (t = l,m),$$

where $p = (p,q) \in T_0$ and $l = (l,m) \in T'$. Then the mapping from T' onto the curved element T can be considered as a non-linear pertubation of the linear transformation. If $x = (x,y) \in T$, we write the complete transformation as

$$x = F(p) = F_0(p) + F_1(p),$$

where F_0 is the linear transformation and F_1 is the non-linear pertubation term. The curved elements considered in Chapter 4 can all be considered in this way. Note that the *construction* of some of the elements in Section 4.6 uses a different sequence of transformations in which the intermediate element T' has curved sides and the same vertices as the standard triangle T_0.

In order that the bounds such as (5.6) can be used to derive estimates of the orders of convergence of individual finite element approximations it is often necessary to map $\mathscr{H}_2^{(r)}(T_0)$ onto $\mathscr{H}_2^{(r)}(T)$ and back again. Thus it is assumed that the transformation F is sufficiently differentiable to ensure that when $v \in \mathscr{H}_2^{(r)}(T)$, it follows that $v \circ F \in \mathscr{H}_2^{(r)}(T_0)$, where the *compound* (or *composite*) *operator* $v \circ F$ is defined by

$$(v \circ F)(p) = v(F(p)) \quad (\text{for all } p \in T_0).$$

To simplify the notation, we frequently write v in place of $v \circ F$, if necessary writing $v(p)$ or $v(x)$ as appropriate, to avoid ambiguity. We also assume that the inverse transformation F^{-1} is sufficiently differentiable, thus when $v(p) \in \mathscr{H}_2^{(r)}(T_0)$ it follows that $v(x) \in \mathscr{H}_2^{(r)}(T)$.

Hypothesis 5.1 (*The regularity hypothesis.*) *If $v \in \mathscr{H}_2^{(r)}(T)$ and the diameter† of the element T is h, then it follows that there exist constants*

---

†The diameter of a triangular element is the length of the longest side, the diameter of a quadrilateral element is the length of the longer diagonal.

C_0, C_1 and C_2 such that

$$| v |_{r, T_0} \leqslant C_1 \{ \inf_{p \in T_0} J(\mathbf{p}) \}^{-1/2} h^r \parallel v \parallel_{r, T}, \tag{5.7a}$$

$$| v |_{r, T_0} \geqslant C_2 \{ \sup_{p \in T_0} J(\mathbf{p}) \}^{-1/2} h^r | v |_{r, T} \tag{5.7b}$$

and

$$0 < \frac{1}{C_0} \leqslant \left\{ \frac{\sup J(\mathbf{p})}{\inf J(\mathbf{p})} \right\} \leqslant C_0, \tag{5.7c}$$

where J is the Jacobian of the transformation $\mathbf{x} = \mathbf{F}(\mathbf{p})$.

Note that this hypothesis implies that the Jacobian is positive for all $\mathbf{p} \in T_0$. In Chapter 4, it is shown that the Jacobian always satisfies this condition if the so-called forbidden elements are avoided.

When the linear transformation $\mathbf{x} = \mathbf{F}_0(\mathbf{p})$ is used, the Jacobian is constant and

$$J_0 = \det \begin{bmatrix} \dfrac{\partial x}{\partial p} & \dfrac{\partial x}{\partial q} \\ \dfrac{\partial y}{\partial p} & \dfrac{\partial y}{\partial q} \end{bmatrix} = \det \begin{bmatrix} 1 & x_1 & y_1 \\ 1 & x_2 & y_2 \\ 1 & x_3 & y_3 \end{bmatrix},$$

the bound (5.7c) is thus trivially satisfied.

The hypothesis has been verified for certain non-linear transformations of the form

$$\mathbf{x} = \mathbf{F}_0(\mathbf{p}) + \mathbf{F}_1(\mathbf{p}),$$

where $\mathbf{F}_1$ is a 'small' pertubation (see Exercises 9 and 10). If sufficient conditions are imposed on the pertubation term $\mathbf{F}_1$, it is possible to show that the Jacobian is

$$J(\mathbf{p}) = J_0 + J_1(\mathbf{p}),$$

where J_1 is also 'small', and hence verify Hypothesis 5.1 (Ciarlet and Raviart, 1972c; and Zlámal, 1974).

Exercise 4   Prove that it follows from (5.7b) that

$$\parallel v \parallel_{r, T_0} \geqslant C \parallel v \parallel_{r, T} h^r \{ \sup_{p \in T_0} J(\mathbf{p}) \}^{-1/2}.$$

Exercise 5   Prove that for any $v \in \mathscr{L}_2(T)$

$$\parallel Jv \parallel_{\mathscr{L}_2(T_0)} \leqslant \{ \sup_{p \in T_0} J(\mathbf{p}) \}^{1/2} \parallel v \parallel_{\mathscr{L}_2(T)}.$$

The Bramble—Hilbert lemma combined with the regularity hypothesis is used primarily but not exclusively to estimate interpolation errors, it can be used equally well to derive bounds on the errors resulting from using numerical quadrature schemes based on finite element partitions of the region of integration.

Exercise 6   (i) Let $E_0(v)$ be the error in integrating $v \in \mathscr{H}_2^{(k+1)}(T_0)$ numerically over the standard element. If the quadrature rule is exact for polynomials of degree not exceeding k, verify that the Bramble–Hilbert lemma leads to the error bound

$$| E_0(v) | \leqslant C \, | \, v \, |_{k+1,T_0} .$$

(ii) Using the regularity hypothesis, verify that if the same rule is transformed onto an arbitrary element using a linear transformation, the error in integrating $v \in \mathscr{H}_2^{(k+1)}(T)$ can be written as $E_0(J_0 v)$ and that

$$| E_0(J_0 v) | \leqslant C h^{k+1} \, | \, v \, |_{k+1,T} .$$

Exercise 7   (i) Let $u \in \mathscr{H}_2^{(k+1)}(T_0)$ and $w \in P_r$ and denote by $E_0(uw)$ the error in integrating the product numerically over the standard element. By introducing the bilinear form $E_1 \in \mathscr{L}(\mathscr{H}_2^{(k+1)}(T_0) \times \mathscr{L}_2(T_0); \mathbb{R})$ defined by

$$E_1(u,w) = E_0(uw),$$

prove that if the quadrature rule is exact for polynomials of degree not exceeding $r + k$, then there is a constant C such that

$$| E_0(uw) | \leqslant C \, \| \, w \, \|_{\mathscr{L}_2(T_0)} \, | \, u \, |_{k+1,T_0} .$$

(Hint. The proof is analogous to the derivation of (5.6).)

(ii) Verify that if the quadrature rule is transformed onto an arbitrary element using a linear transformation, then the error in integrating the product of $u \in \mathscr{H}_2^{(k+1)}(T)$ and $w \in P_r$ numerically can be written as $E_0(J_0 uw)$ and that

$$| E_0(J_0 uw) | \leqslant C h^{k+1} \, | \, u \, |_{k+1,T} \, \| \, w \, \|_{\mathscr{L}_2(T)} .$$

In Exercises 6 and 7 it is assumed that the transformations are linear as this ensures that when $w(\mathbf{x}) \in P_r$ then $J_0 w(\mathbf{p}) \in P_r$. If a non-linear transformation is used as in Section 5.4(A), it is necessary to consider functions $w(\mathbf{x})$ such that $w(\mathbf{p})J(\mathbf{p})$ is a polynomial. Further details of the transformation of quadrature rules onto arbitrary elements can be found in Section 5.4(A).

Exercise 8   Prove that if $\mathbf{x} = \mathbf{F}_0(\mathbf{p})$, then there exist constants C_1 and C_2 such that

$$\iint_{T_0} \left(\frac{\partial v}{\partial t}\right)^2 d\mathbf{p} \leqslant C_1 h^2 \iint_T \left\{ \left(\frac{\partial v}{\partial x}\right)^2 + \left(\frac{\partial v}{\partial y}\right)^2 \right\} J_0 d\mathbf{x} \quad (t = p,q)$$

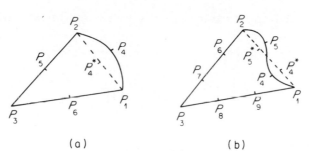

Figure 24

and

$$\iint_{T_0} \left(\frac{\partial v}{\partial t}\right)^2 dp \geq C_2 h^2 \iint_T \left\{\left(\frac{\partial v}{\partial x}\right)^2 + \left(\frac{\partial v}{\partial y}\right)^2\right\} J_0 dx \quad (t = p,q).$$

Hence prove by induction that (5.7a) and (5.7b) are valid for $r = 1,2, \ldots$ with $\| v \|_{r,T}$ replaced by $| v |_{r,T}$.

Exercise 9 (a) Verify that if P_5 and P_6 are the mid-points of the sides $P_2 P_3$ and $P_1 P_3$ respectively, then quadratic isoparametric elements with one curved side (see Figure 24(a)), lead to the trans-formation

$$t = t_3 + (t_1 - t_3)p + (t_2 - t_3)q + \left(t_4 - \frac{t_1 + t_2}{2}\right) 4pq \quad (l = x,y).$$

(cf. 4.6).

(b) Verify that for such a transformation

$$J(\mathbf{p}) = J_0 + J_1 (\mathbf{p}),$$

where

$$J_1 (\mathbf{p}) = 4 \left(x_4 - \frac{x_1 + x_2}{2}\right) \{(y_2 - y_3)q - (y_1 - y_3)p\}$$

$$+ 4 \left(y_4 - \frac{y_1 + y_2}{2}\right) \{(x_1 - x_3)p - (x_2 - x_3)q\}.$$

Find sufficient conditions to ensure that

$$0 < J_0 - Ch^3 \leq J \leq J_0 + Ch^3$$

and hence that $J = 0(h^2)$. Compare your results with (5.19).

Zlámal (1973) has provided an alternative formulation for triangular elements with one curved side that leads to similar estimates.

Exercise 10   Verify that if P_6, P_7, P_8 and P_9 are the points of trisection of the sides $P_2 P_3$ and $P_1 P_3$ (see Figure 24(b)), then the cubic isoparametric elements lead to a transformation with non-linear terms;

$$27pq(1-p-q)\left(t_{10}-\frac{t_1+t_2+t_3}{3}\right)+\frac{9}{2}pq\left[(3p-1)\left(t_4-\frac{2t_1+t_2}{3}\right)\right.$$

$$\left.+(3q-1)\left(t_5-\frac{2t_2+t_1}{3}\right)\right]\qquad(t=x,y).$$

Complete approximations

The regularity hypothesis contains assumptions on the properties of the transformation from an arbitrary element to the standard element. When these assumptions are combined with the Bramble–Hilbert lemma it is possible to derive bounds such as (5.4) and hence determine the order of convergence of a finite element approximation. Thus the primary function of the Bramble–Hilbert lemma is to provide bounds on errors in interpolation. The lemma can be used equally well to bound the error in any form of approximation that is a projection onto a space of (piecewise) polynomials. In order to apply the lemma it is first necessary to make some assumptions concerning the types of function that makes up the finite element approximation.

For any element T, we denote by $K_{[T]}$ the space defined by trial functions that can be non-zero in T; that is, $K_{[T]}$ is the restriction of the space of trial functions to the element T. We denote by $K_{[0]}$, the space of functions $v(\mathbf{p})$ such that $v(\mathbf{x}) \in K_{[T]}$. It is this space $K_{[0]}$ that has a critical significance in the analysis of finite element methods. The order of the method is governed by the maximum degree of polynomial (in $\mathbf{p}$) which, when interpolated by a function in $K_{[0]}$, leads to a zero interpolation error. In general this is equivalent to the maximum value of k such that $P_k \subset K_{[0]}$.

For example, if the approximation is in terms of piecewise cubic polynomials (Lagrange or Hermite) on a triangular mesh (Section 4.1), then the transformation $\mathbf{F}$ is linear and $K_{[0]}, K_{[T]} = P_3$. Whereas the eighteen-parameter C^1-quintics (4.14) lead to a linear transformation, $P_4 \subset K_{[0]}, K_{[T]} \subset P_5$ (strict inclusion); while if the biquadratic isoparametric approximation (4.29) is used, it follows that $P_2 \subset K_{[0]} \subset P_4$ (again strict inclusion) but that as the transformation is non-linear, $K_{[T]}$ is not a polynomial subspace.

For any element T, we introduce the projection Π_T onto $K_{[T]}$, that is, for any sufficiently differentiable function u, $\Pi_T u(\mathbf{x})$ interpolates $u(\mathbf{x})$ in T. Interpolation in this context means matching all the nodal parameters used in a finite element approximation. For functions defined on the standard element it is possible to define a mapping Π onto $K_{[0]}$ by

$$\Pi(v \circ \mathbf{F}) = (\Pi_T v) \circ \mathbf{F}.$$

Hypothesis 5.2   (The completeness hypothesis.) If T is any element of diameter h, then there exists $k > 0$ such that $P_k \subset K_{[0]}$ and Π is a projection for which $\| I - \Pi \|$ is uniformly bounded for all h.

An example in which $\| I - \Pi \|$ is not uniformly bounded occurs using triangular elements when the normal derivative at a side point is a parameter but the function value and tangential derivative are not, and triangles tend to degenerate as the mesh is reduced, that is, there are elements in which the smallest angle tends to zero (Bramble and Zlámal, 1970). This example is mentioned again towards the end of Section 5.3.

In Section 5.3, it is assumed that it is possible to define $\mathscr{H}_2^{(k)}(R)$ for non-integer values of k, as well as for integers. A detailed discussion of such a space and the significance of the *trace theorem* is beyond the scope of this book and the interested reader is referred to Aubin (1972), Lions and Magenes (1972) or Nečas (1967). A brief account of some of the more important properties and their uses can be found in Chapter 1 of Varga (1971).

5.2 CONVERGENCE OF GALERKIN APPROXIMATIONS

The results of Section 3.5 are extended to show that, in general, a Galerkin approximation for linear problems is *near best* in the sense described in the previous section. The basic result is an extension of the *Lax—Milgram lemma* (Yosida, 1965, p. 92):

Theorem 5.2   (Aubin, 1972, p. 34) If $\mathscr{H}$ is a Hilbert space, $l \in \mathscr{H}'$ and a is an $\mathscr{H}$-elliptic and bounded bilinear form, then there exists a unique $u \in \mathscr{H}$ such that

$$a(u,v) = l(v) \quad \text{(for all } v \in \mathscr{H} \text{)}.$$

Given an N-dimensional subspace K_N, there exists a unique $U \in K_N$ such that

$$a(U,V) = l(V) \quad \text{(for all } V \in K_N \text{)}.$$

Furthermore

$$\| u - U \|_{\mathscr{H}} \leqslant C \inf_{\tilde{u} \in K_N} \| u - \tilde{u} \|_{\mathscr{H}}.$$

A similar result can be obtained for certain non-linear problems using monotone operator theory (Varga, 1971, Chapter 4).

As an example of the results that can be derived from Theorem 5.2, consider a Ritz approximation to the solution of

$$\frac{\partial}{\partial x}\left(d_1(x,y)\frac{\partial u}{\partial x}\right) + \frac{\partial}{\partial y}\left(d_2(x,y)\frac{\partial u}{\partial y}\right) + f(x,y) = 0 \quad ((x,y) \in R)$$

$$(5.8)$$

subject to

$$u(x,y) = 0 \quad ((x,y) \in \partial R),$$
(5.9)

where there exist constants δ and Δ such that

$$0 < \delta \leqslant d_1(x,y), d_2(x,y) \leqslant \Delta \quad ((x,y) \in R).$$

In this application

$$l(v) = (f,v)$$

and

$$a(u,v) = \iint_R \left\{ d_1 \frac{\partial u}{\partial x} \frac{\partial v}{\partial x} + d_2 \frac{\partial u}{\partial y} \frac{\partial v}{\partial y} \right\} dx \, dy.$$

Since

$$| a(u,v) | \leqslant \Delta \iint_R \left\{ \left| \frac{\partial u}{\partial x} \frac{\partial v}{\partial x} \right| + \left| \frac{\partial u}{\partial y} \frac{\partial v}{\partial y} \right| \right\} dx dy,$$

it is bounded and since

$$a(u,u) \geqslant \delta \, | u |_{1,R},$$

it follows from Lemma 5.1 that it is $\overset{\circ}{\mathscr{H}}_2^{(1)}(R)$-elliptic.

Corollary of Theorem 5.2   *If u is the solution of (5.8) and (5.9), then the Ritz approximation $U \in \overset{\circ}{K}_N \subset \overset{\circ}{\mathscr{H}}_2^{(1)}(R)$ is such that*

$$\| u - U \|_{1,R} \leqslant C \inf_{\tilde{u} \in \overset{\circ}{K}_N} \| u - \tilde{u} \|_{1,R}.$$

Exercise 11   Prove that if (i) u is the solution of (5.8) subject to $u = g$ on ∂R and (ii) $w \in \mathscr{H}_2^{(1)}(R)$ is *any* function such that $w = g$ on ∂R then the Ritz approximation $U + w$ with $U \in \overset{\circ}{K}_N \subset \overset{\circ}{\mathscr{H}}_2^{(1)}(R)$ is such that

$$\| u - (U + w) \|_{1,R} \leqslant C \inf_{\tilde{u} \in \overset{\circ}{K}_N} \| u - (\tilde{u} + w) \|_{1,R}$$

Exercise 12   Prove that, if u is a solution of (5.8), subject to $\partial u/\partial n = g$ on ∂R, then there exists a Ritz approximation $U \in K_N \subset \mathscr{H}_2^{(1)}(R)$ such that

$$\| u - U \|_{1,R} \leqslant C \inf_{\tilde{u} \in K_N} \| u - \tilde{u} \|_{1,R}.$$

Note that it is necessary to assume a *compatability condition* on f and g if the solution is to exist. For example, if $d_1 = d_2 = 1$ in (5.8), then we assume that

$$\iint_R f \, dx \, dy = - \int_{\partial R} g \, d\sigma$$

and as the solution is only unique to within an additive constant, it is possible to choose a constant such that

$$\iint_R u \, dx \, dy = 0$$

(Nečas, 1967, p. 256). Hence it is possible to apply Lemma 5.3 with $k = 1$, to obtain the result.

If it is necessary to approximate the boundary conditions in terms of basis functions that take non-zero values on the boundary, it is still possible to derive error bounds. If

$$U = U_0 + \bar{U}$$

where $U_0 \in \mathring{K}_N \subset \mathring{\mathscr{H}}_2^{(1)}(R)$ then $\bar{U}$ contains *only* basis functions that can take non-zero values on the boundary and is completely determined by the boundary data. It follows from the definitions in Section 5.1 that $\bar{U} \in \bar{K}_N$.

Lemma 5.4 *If u is the solution of (5.8), subject to $u = g$ on ∂R and $\bar{U} \in \bar{K}_N$ is chosen such that $\bar{U}(x,y)$ $((x,y) \in \partial R)$ is a fixed approximation to g, then the finite element approximation $U = U_0 + \bar{U}$ such that $U_0 \in \mathring{K}_N \subset \mathring{\mathscr{H}}_2^{(1)}(R)$ is such that*

$$|u - U|_{1,R} \leqslant C \, | \, u - (\tilde{u} + \bar{U}) \, |_{1,R} \quad \text{(for all } \tilde{u} \in K_N\text{)}.$$

Proof Since

$$u - (\tilde{u} + \bar{U}) = u - U + U_0 - \tilde{u}$$

and $U_0 - \tilde{u} \in \mathring{K}_N$ the result follows directly from Lemma 5.3 and the definitions of u and U.

It is also possible to use Lemma 5.4 to derive bounds in terms of $\mathscr{H}_2^{(1)}(R)$ if we assume that there exists a smooth continuation of g into R. At least one such continuation exists, namely the solution u itself.

Theorem 5.3 (Fairweather, 1972, p. 45) *Let u be the solution of (5.8), subject to $u = g$ on ∂R and let $w \in \mathscr{H}_2^{(1)}(R)$ be any smooth continuation of g into R. Then the finite element approximation U, which can be written as $U = U_1 + W$, where $W \in K_N$ is an approximation to w and $U_1 \in \mathring{K}_N$, is such that*

$$\| u - (U_1 + W) \|_{1,R} \leqslant C\{ \| u - (\tilde{u} + W) \|_{1,R} + \| w - W \|_{1,R} \}$$

$$(5.10)$$

for any $u \in \mathring{K}_N$.

Note that if, for example, the boundary condition is interpolated, the right-hand side of (5.10) consists of the error in approximating $u - w \in \mathring{\mathscr{H}}_2^{(1)}(R)$ together with the error in interpolating the function w, which is non-zero on the boundary.

Boundary approximation and quadrature

Computing a finite element approximation with interpolated boundary conditions — as above — is only one of the major *variational crimes* (Strang, 1972) invariably committed when solving practical problems. The others are (i) altering the position of the boundary, (ii) using numerical integration (quadrature) for the inner products and (iii) using non-conforming elements. Non-conforming elements are also considered in detail in Section 7.2. If any of these techniques are used then it follows that the approximate solution is no longer in K_N nor satisfies

$$a(U,V) = (f,V) \quad \text{(for all } V \in K_N\text{).}} \tag{5.11}$$

It is instead $U_h \in K_h$ and satisfies

$$a_h(U_h,V_h) = (f,V_h)_h \quad \text{(for all } V_h \in K_h\text{),}} \tag{5.12}$$

where both sides of (5.12) and the form of the new M-dimensional space K_h, which may contain functions that are not admissible for the classical variational methods outlined in Chapter 3, depend on the nature of the variational crimes committed. It is usually the case that $M \geqslant N$ and that $K_h \supseteq K_N$. If $K_h = K_N$, as when the use of quadrature is the only deviation from the classical form of the variational method, the solutions of both (5.11) and (5.12) are linear combinations of $\varphi_i(x)$ ($i = 1, \ldots, N$). The coefficients of the solution to (5.11) are given by

$$G\alpha = b, \tag{5.13}$$

where the *stiffness matrix* $G = \{a(\varphi_i, \varphi_j)\}$ ($i = 1, \ldots, N$) and $b = \{(f, \varphi_i)\}$. The perturbed problem (5.12) leads to

$$G_h \alpha_h = b_h, \tag{5.14}$$

where $G_h = \{a_h(\varphi_i, \varphi_j)\}$ and $b_h = \{(f, \varphi_i)_h\}$. It is possible to compare the individual elements of G_h and b_h with the corresponding elements of G and b, then derive bounds on $\| U - U_h \|$ by a standard perturbation analysis of linear algebra. Unfortunately, this method is known to lead to a poor upper bound on $\| U - U_h \|$ in most cases (Fix, 1972). If it is assumed that the perturbed form is $\mathscr{H}$-elliptic and bounded then not only is the existence of a unique solution guaranteed by Theorem 5.2, but it is also possible to estimate the size of the perturbation.

Exercise 13   Verify that if a_h is an $\mathscr{H}$-elliptic bilinear form then

$$a_h(U - U_h, W_h) = (a_h - a)(U, W_h) + (f, W_h) - (f, W_h)_h,$$

for any $W_h \in K_h$. Hence prove that

$$\| U - U_h \| \leqslant C \sup_{W_h \in K_N} \left\{ \frac{| (a_h - a)(U, W_h) | + | (f, W_h) - (f, W_h)_h |}{\| W_h \|} \right\} \tag{5.15}$$

and construct a similar bound for $\| u - U_h \|_{\mathscr{H}}$.

Exercise 14   Verify that if a_h is a bounded bilinear form then

$$|a_h(U_h - V_h, W_h)| \leq \gamma \|u - V_h\| \|W_h\| + |(f, W_h)_h - a_h(u, W_h)|$$

for any $V_h, W_h \in K_h$. Hence prove that if a_h is also $\mathscr{H}$-elliptic

$$\|U_h - V_h\| \leq \frac{\gamma}{\alpha} \|u - V_h\| + \sup_{W_h \in K_h} \left\{ \frac{1}{\alpha} \frac{|(f, W_h)_h - a_h(u, W_h)|}{\|W_h\|} \right\}$$

for any $V_h \in K_h$, and that there exists $C > 0$ such that

$$\|u - U_h\| \leq C \left\{ \inf_{V_h \in K_h} \|u - V_h\| \right.$$

$$\left. + \sup_{W_h \in K_h} \left\{ \frac{|(f, W_h)_h - a_h(u, W_h)|}{\|W_h\|} \right\} \right\}. \quad (5.16)$$

Note that the estimate (5.15) can only be used when U_h is in some sense a perturbation of a classical Galerkin approximation. That is, if $K_h = K_N$ or $K_h = K_N \oplus \{\varphi_{N+1}, \ldots, \varphi_M\}$, where the additional $\varphi_i(x)$ $(i = N + 1, \ldots, M)$ have special properties that exclude them as admissible functions for a classical approximation; they could be non-conforming for example. Alternatively the estimate (5.16) is of use when the approximation differs significantly from any classical approximation such as when non-conforming elements alone are used (Section 5.4(E)).

The same form of perturbation analysis can be used when (5.5) represents a system of *variational difference equations* (for example, Dem'janovič, 1964).

Exercise 15   Prove that if a_h and a are bilinear forms such that a_h is K_h-elliptic and a is bounded then

$$\|u - U_h\| \leq C \left\{ \inf_{V_h \in K_h} \left[\|u - V_h\| + \sup_{W_h \in K_h} \left\{ \frac{|(a_h - a)(V_h, W_h)|}{\|W_h\|} \right\} \right] \right.$$

$$\left. + \sup_{W_h \in K_h} \left\{ \frac{|(f, W_h)_h - (f, W_h)|}{\|W_h\|} \right\} \right\}.$$

$$(5.17)$$

5.3 APPROXIMATION ERRORS

As is shown in the previous section, Galerkin approximations are invariably *near best* in some norm. In particular, approximate solutions to second-order problems are such that

$$\|u - U\|_{1, R} \leq C \inf_{\tilde{u} \in K_N} \|u - \tilde{u}\|_{1, R}.$$

Bounds on the error have also been derived in terms of other norms, in particular the $\mathscr{L}_2(R)$ norm and the $\mathscr{L}_\infty(\bar{R})$ norm. Estimates in $\mathscr{L}_\infty(\bar{R})$ have also been used to prove the so-called *superconvergence* properties of certain Galerkin methods; that is, the approximation at the nodes is of a higher order than at non-nodal points (Douglas, Dupont and Wheeler, 1974). Further reading can be found in de Boor (1974).

Let T_0 be the *standard element*, then most of the convergence results can be derived from the following lemma concerning polynomial approximation on T_0:

Lemma 5.5 Let $\Pi \in \mathscr{L}(\mathscr{H}_2^{(k+1)}(T_0); \mathscr{H}_2^{(r)}(T_0))$ $(k \geqslant r)$ *be a projection onto* $K_{[0]}$ *where* $\mathscr{H}_2^{(r)}(T_0) \supset K_{[0]} \supset P_k$. *Then*

$$\| v - \Pi v \|_{r, T_0} \leqslant C \| I - \Pi \| \, | v |_{k+1, T_0}$$

for all $v \in \mathscr{H}_2^{(k+1)}(T_0)$.

That is, if the interpolation is exact for polynomials of degree not exceeding k, the error in interpolation can be expressed in terms of the $(k+1)$th derivatives of the function being interpolated. The operators Π and $I - \Pi$ are defined from $\mathscr{H}_2^{(k+1)}(T_0)$ onto $\mathscr{H}_2^{(r)}(T_0)$ because when this result is applied later, r depends on the order of the differential equation and k depends on the properties of the trial functions. The value of $\| I - \Pi \|$ depends on the values of r and k, but it is assumed to be uniformly bounded and the precise value is not, in general, important.

Proof of Lemma 5.   For any $G \in \mathscr{H}_2^{(-r)}(T_0)$ define $F \in \mathscr{H}_2^{(-k-1)}(T_0)$ such that for any $v \in \mathscr{H}_2^{(k+1)}(T_0)$

$$F(v) = G([I - \Pi]v).$$

Then by applying the Bramble–Hilbert lemma to F, we have that

$$| G([I - \Pi]v) | \leqslant C \| F \|_{-k-1, T_0} \, | v |_{k+1, T_0},$$

where

$$\| F \|_{-k-1, T_0} = \sup_{w \in \mathscr{H}_2^{(k+1)}(T_0)} \left\{ \frac{| G([I - \Pi]w) |}{\| w \|_{k+1, T_0}} \right\}.$$

It follows from the duality of $\mathscr{H}_2^{(r)}(T_0)$ and $\mathscr{H}_2^{(-r)}(T_0)$ that for any $u \in \mathscr{H}_2^{(r)}(T_0)$

$$\| u \|_{r, T_0} = \sup_{G \in \mathscr{H}_2^{(-r)}(T_0)} \left\{ \frac{| G(u) |}{\| G \|} \right\},$$

thus in particular

$$\| (I - \Pi)v \|_{r, T_0} = \sup_G \left\{ \frac{| G([I - \Pi]v) |}{\| G \|} \right\}.$$

Then combining these results, it follows that

$$\| (I - \Pi)v \|_{r, T_0} \leqslant C \mid v \mid_{k+1, T_0} \sup_G \left\{ \frac{\| F \|}{\| G \|} \right\}$$

$$\leqslant C \mid v \mid_{k+1, T_0} \sup_G \left\{ \sup_{w \in \mathscr{H}_2^{(k+1)}(T_0)} \left\{ \frac{\mid G([I - \Pi]w) \mid}{\| w \|_{k+1, T_0}} \right\} \frac{1}{\| G \|} \right\},$$

but

$$\mid G([I - \Pi]w) \mid \leqslant \| G \| \, \| I - \Pi \| \, \| w \|_{k+1, T_0}$$

and so the result follows immediately.

This result can be combined with the regularity hypothesis to provide an interpolation error bound in the region R.

Theorem 5.4   Assume that the completeness hypothesis is valid for some $k > 0$ and that the regularity hypothesis is valid for all $r \leqslant k$. Then if $u \in \mathscr{H}_2^{(k+1)}(R)$ and u interpolates u, it follows that

$$\| u - \tilde{u} \|_{r, R} \leqslant Ch^{k+1-r} \mid u \mid_{k+1, R},$$

where h is a bound on the diameters of the elements of the partition of R.

Proof.   If we assume that the region R is partitioned into elements T_j $(j = 1, \ldots, S)$ then

$$\| u - \tilde{u} \|_{r, R}^2 = \sum_{j=1}^{S} \| u - \Pi_{T_j} u \|_{r, T_j}^2.$$

Considering a typical element T, it follows from Lemma 5.5 that,

$$\| u \quad \Pi u \|_{r, T_0} \leqslant C \| I - \Pi \| \mid u \mid_{k+1, T_0}$$

and then by Hypothesis 5.1 and Exercise 4,

$$\| u \quad \Pi_T u \|_{r, T} \leqslant Ch^{k+1-r} \| I - \Pi \| \mid u \mid_{k+1, T}.$$

As there exists a uniform bound on all the operator norms $\| I - \Pi \|$, summing over all elements leads to the desired result.

One particular situation in which this result can be applied is when the transformation T_0 to T is linear and interpolation by finite element approximating functions is exact for $u \in P_k$. Then in particular

$$\| u - \tilde{u} \|_{1, R} \leqslant Ch^k \mid u \mid_{k+1, R},$$

where $\tilde{u}$ is the interpolating function. This is the type of bound required to estimate the order of the finite element approximation for second-order problems using the near-best inequalities of Section 5.2. Note that if we consider a region made up of one element we have an error bound

120

for classical Lagrange interpolation; this will be used later, in Section 5.4(B).

The Lagrangian (or Hermitian) elements of Section 4.1 are such that the transformation from T_0 to T is linear, the basis functions $\varphi \in P_k$, interpolation is exact for $u \in P_k$ and $K_{[0]} = K_{[T]} = P_k$. Thus if the region R is a polygon with the boundary conditions matched exactly and the integrals evaluated analytically, the error in such a finite element approximation to a second-order problem is such that

$$\| u - U \|_{1,R} \leqslant Ch^k \, |u|_{k+1,R}. \tag{5.18}$$

An equivalent result holds for problems on rectangles with a rectangular grid; again it is the degree of polynomials that can be interpolated exactly that governs the exponent of h in the error bound (5.18).

Exercise 16   Verify that eliminating the internal parameters from finite element approximations such as cubic Lagrangian (or Hermitian) elements, or eliminating the normal derivatives at the side mid-points of the 21-parameter C^1-quintics, leads to a reduction of one in the exponent of h in (5.18).

Exercise 17   Verify that subparametric biquadratic and bicubic approximation (Section 4.3), using a bilinear transformation from an arbitrary quadrilateral element onto the unit square, can interpolate exactly quadratic and cubic polynomials respectively, in *x and y*. Show that if R is a polygon, the boundary conditions are matched exactly and the integrals are evaluated analytically, then the error bound (5.18) is valid for $k = 2$ and 3 respectively.

Exercise 18   Verify that (5.18) holds for hexahedral subparametric approximations in three dimensions, provided that the region R can be partitioned exactly.

Curved elements

Ciarlet and Raviart (1972b) have shown that error bounds such as (5.18) are also valid for certain isoparametric elements with a *single curved side*; in particular they verified the order of convergence for two special cases:

(1) *Quadratic isoparametric triangular elements* can be written (Exercise 9) as

$$t = t_3 + (t_1 - t_3)p + (t_2 - t_3)q + \left(t_4 - \frac{t_1 + t_2}{2} \right)4pq \quad (t = x,y).$$

Since the interpolation is exact for quadratic polynomials in p and q, the bound on the error is

$$\| u - U \|_{1,R} = O(h^2),$$

provided that

$$\left\{ \left[x_4 - \frac{x_1 + x_2}{2} \right]^2 + \left[y_4 - \frac{y_1 + y_2}{2} \right]^2 \right\}^{1/2} = O(h^2). \tag{5.19}$$

This is equivalent to

$$\| P_4 - P_4^* \|_{\mathbb{R}^2} = O(h^2),$$

where P_4^* is the mid-point of the *chord* $P_1 P_2$ and the norm is the Euclidean distance in $\mathbb{R}^2$. This additional condition can always be satisfied if h is sufficiently small compared to the radius of curvature of the boundary. It is also assumed that the boundary can be represented *exactly* in terms of arcs that can be parametrized in the form

$$t = t_1 p(2p - 1) + t_2(1 - p)(1 - 2p) + t_4 4p(1 - p) \tag{5.20}$$

$$(t = x, y; p \in [0,1]).$$

A similar bound holds for biquadratic isoparametric approximations based on quadrilaterals with a single curved side, if similar conditions are satisfied.

(2) *Cubic isoparametric elements* can be written (Exercise 10) as

$$t = t_3 + (t_1 - t_3)p + (t_2 - t_3)q$$
$$+ 27pq(1 - p - q) \left(t_{10} - \frac{t_1 + t_2 + t_3}{3} \right)$$
$$+ \tfrac{9}{2} pq \left[(3p - 1) \left(t_4 - \frac{2t_1 + t_2}{3} \right) \right.$$
$$\left. + (3q - 1) \left(t_5 - \frac{2t_2 + t_1}{3} \right) \right] \quad (t = x, y).$$

Since the interpolation is exact for cubic polynomials in p and q, the bound on the error is then

$$\| u - U \|_{1,R} = O(h^3),$$

provided that

$$\| P_j - P_j^* \|_{\mathbb{R}^2} = O(h^2) \quad (j = 4, 5)$$

and

$$\| (P_4 - P_4^*) - (P_5 - P_5^*) \|_{\mathbb{R}^2} = O(h^3), \tag{5.21}$$

and that the point $P_{1\,0}$ is selected such that

$$t_{1\,0} = \frac{t_1 + t_2 + t_3}{3} + \frac{(t_4 - t_4^*) + (t_5 - t_5^*)}{4} \qquad (t = x,y), \qquad (5.22)$$

where $P_4^* = (x_4^*, y_4^*)$ and $P_5^* = (x_5^*, y_5^*)$ are the points of trisection of the *chord* $P_1 P_2$ adjacent to P_4 and P_5 respectively. The additional inequality (5.21) is in fact a realistic condition if h is sufficiently small. As for the quadratic case, the bound on the finite element error is only valid if there is *no boundary perturbation*. A similar bound holds for Hermitian isoparametric approximation (Ciarlet and Raviart, 1972b), subject to a set of conditions analogous to (5.21) and (5.22).

These severe conditions on the curvature of isoparametric elements may be necessary in practice, as certain numerical evidence would indicate (Bond, Swannell, Henshell and Warburton, 1973) — but opinions differ.

In all the preceding convergence estimates in two dimensions it is assumed that θ, the smallest angle subtended by a triangular mesh at any node, is bounded away from zero as h tends to zero. In the analogous results for three dimensions, or for quadrilateral meshes in two dimensions, it is assumed that the ratio h/ρ remains bounded as h tends to zero. For any element, ρ is the diameter of the largest sphere (in $\mathbb{R}^3$) — circle in $\mathbb{R}^2$ — that is contained in the element. If it is not possible to make such assumptions then the interpolation error bounds take the form

$$\| u - \tilde{u} \|_{r,R} = O\left(\frac{h^{k+1}}{\rho^r}\right), \qquad (5.23)$$

assuming that $\| I - \Pi \|$ is uniformly bounded for all elements. This modified bound is introduced because in Hypothesis 5.1 the second inequality (5.7b) is now in terms of ρ rather than h (Ciarlet and Raviart, 1972a; Bramble and Zlámal, 1970). For triangular meshes

$$\rho \approx h \sin \theta,$$

hence the bound can be written in terms of $h^{k+1-r}/(\sin \theta)^r$. If it is not assumed that h/ρ is bounded as h tends to zero, then examples exist† for which $\| I - \Pi \|$ is not uniformily bounded and hence, as is shown by Bramble and Zlámal, the interpolation error bound can take the form

$$\| u - \tilde{u} \|_{r,R} \leqslant C \frac{h^{k+1-r}}{(\sin \theta)^{n+r}} | u |_{k+1,R}$$

†The 21-parameter C^1-quintic approximation is one such example.

for some $n \geqslant 1$, rather than (5.18). Babuška and Aziz (1976) have indicated that it is better to consider angles that tend to 2π rather than those that tend to zero.

Interpolation error bounds of the form

$$\| u - \tilde{u} \|_{r,R} \leqslant C_{k,r} h^{k+1-r}, \tag{5.24}$$

have been derived using the *Sard kernel theorem* from methods based on both rectangles and triangles. For some forms of piecewise polynomial approximations, numerical values have been computed for the constants $C_{k,r}$ in (5.24). These results also suggest that bounds of the form (5.23) are not the best possible (Barnhill, Gregory and Whiteman, 1972, and references therein).

Interpolation error bounds in terms of the maximum semi-norm $|\cdot|_{(k)\bar{R}}$ rather than the Sobolev semi-norm as in (5.18), were first derived by Zlámal and later by Ciarlet and Raviart and Zeníšek (Ciarlet and Raviart, 1972a, and references therein).

If the solution is not sufficiently smooth $|u|_{k+1,R}$ may not exist and it will not be possible to derive a bound such as (5.18), even though $K_{[0]} \supset P_k$. In such cases, it is necessary to use the exponent

$$k^* = \max\{s : u \in \mathscr{H}_2^{(s)}(R)\};$$

since $k^* \leqslant k$, the only modification of Theorem 5.4 that is necessary is to replace k by k^* wherever it appears, the analysis is unaltered. Typical problems for which the solution is singular, or nearly so, are defined on regions with re-entrant corners. Various special techniques have been developed for such problems and these are considered briefly in Section 7.4(F), an alternative method is given in Chapter 8 of Strang and Fix (1973) and also in Wait (1976).

5.4 PERTURBATION ERRORS

The error bounds (5.16) and (5.17) involve two distinct types of error:

(1) The first term

$$\inf_{V_h \in K_h} \| u - V_h \|$$

is an *approximation error* and it can be bounded by the methods outlined in Section 5.3.

(2) All the remaining terms are introduced because of the perturbed form of the Galerkin equation (5.12). In this section we seek bounds on these perturbation terms for different forms of the perturbation.

The finite element solution is said to be *optimal*, if the order of the perturbation errors are no bigger than that of the approximation error (Nitsche, 1972).

In their analysis of quadrature perturbations, Herbold and Varga (1972) call the quadrature scheme *consistent* if the resulting approximation is optimal.

(A) Numerical integration

Bounds on the errors in integrals evaluated numerically by means of a standard quadrature rule are considered in Exercise 6 and 7 in Section 5.1. In this section we consider such quadratures in more detail and provide bounds that are valid for certain types of non-linear transformation from T_0 onto T.

A quadrature scheme on the standard element is defined as a set of points $p_l \in T_0$ ($l = 1, \ldots, L$) and a set of positive weights b_l ($l = 1, \ldots, L$). The condition $b_l > 0$ is necessary if the perturbed bilinear form a_h is to be K_h-elliptic. Any integral on the standard element can be written as

$$\iint_{T_0} u(\mathbf{p})d\mathbf{p} = \sum_{l=1}^{L} b_l u(\mathbf{p}_l) + E_0(u),$$

where — as before — E_0 is the quadrature error operator for the standard element. To transform the quadrature scheme onto an arbitrary element we take the points $\mathbf{x}_l = \mathbf{F}(\mathbf{p}_l) \in T$ and the weights $\beta_l = b_l J(\mathbf{p}_l)$, where J is the Jacobian of the transformation $\mathbf{F}$. If we denote the quadrature error in T by $E(u)$, it follows that

$$E(u) = \iint_T u(\mathbf{x})d\mathbf{x} - \sum_{l=1}^{L} \beta_l u(\mathbf{x}_l)$$

$$= \iint_{T_0} u(\mathbf{p})J(\mathbf{p})d\mathbf{p} - \sum_{l=1}^{L} b_l J(\mathbf{p}_l)u(\mathbf{p}_l)$$

$$= E_0(uJ).$$

In this section we follow the analysis of Ciarlet and Raviart (1972c) and study the perturbation errors in the solution of the differential equation

$$\frac{\partial}{\partial x}\left(d_1 \frac{\partial u}{\partial x}\right) + \frac{\partial}{\partial y}\left(d_2 \frac{\partial u}{\partial y}\right) + f(x,y) = 0 \quad ((x,y) \in R) \qquad (5.25)$$

subject to

$$u(x,y) = 0 \qquad\qquad ((x,y) \in \partial R).$$

It is assumed that the perturbations are due entirely to evaluating the inner products by numerical quadrature. Thus all the basis functions φ

satisfy the boundary condition and are conforming — in this problem $\varphi \in \mathscr{H}_2^{(1)}(R)$. It is further assumed that the region R is partitioned into S elements T_j ($j = 1, \ldots, S$), and that corresponding to each element there is a transformation F_j from the standard element; the Jacobian of F_j is denoted by J_j.

The bilinear form corresponding to (5.25) is therefore

$$a(u,v) = \sum_{j=1}^{S} \iint_{T_j} \left\{ d_1 \left(\frac{\partial u}{\partial x} \right) \left(\frac{\partial v}{\partial x} \right) + d_2 \left(\frac{\partial u}{\partial y} \right) \left(\frac{\partial v}{\partial y} \right) \right\} dxdy$$

and the perturbed form is

$$a_h(u,v) = \sum_{j=1}^{S} \sum_{l=1}^{L} \beta_{lj} \left\{ d_1 \left(\frac{\partial u}{\partial x} \right) \left(\frac{\partial v}{\partial x} \right) + d_2 \left(\frac{\partial u}{\partial y} \right) \left(\frac{\partial v}{\partial y} \right) \right\}_{x = F_j(\mathbf{p}_l)},$$

where

$$\beta_{lj} = b_l J_j(\mathbf{p}_l) \quad \begin{cases} j = 1, \ldots, S \\ l = 1, \ldots, L. \end{cases}$$

Thus

$$(a - a_h)(u,v) = \sum_{j=1}^{S} \left\{ E_0 \left(d_1 \frac{\partial u}{\partial x} \frac{\partial v}{\partial x} J_j \right) + E_0 \left(d_2 \frac{\partial u}{\partial y} \frac{\partial v}{\partial y} J_j \right) \right\}. \quad (5.26)$$

The individual terms in the summation are thus of the form $E_0(zw)$, where $z(\mathbf{p})$ and $w(\mathbf{p})$ are $d_1(\partial u/\partial x)$ and $(\partial v/\partial x)J_j$ respectively (or $d_2(\partial u/\partial y)$ and $(\partial v/\partial y)J_j$) and we assume that d_1 and d_2 are sufficiently differentiable to ensure that $z \in \mathscr{H}_2^{(k)}(T_0)$. Then it is possible to use the bound on integrals of products derived in Exercise 7, provided $(\partial v/\partial x)J_j$ and $(\partial v/\partial y)J_j$ are polynomials in $\mathbf{p}$.

Similarly the quadrature scheme is applied to the right-hand side of the Galerkin equation

$$a(u,v) = (f,v),$$

where

$$(f,v) = \sum_{j=1}^{S} \iint_{T_j} f(\mathbf{x})v(\mathbf{x})dx.$$

Thus it follows that

$$(f,v)_h = \sum_{j=1}^{S} \sum_{l=1}^{L} \beta_{lj} \{f(\mathbf{x})v(\mathbf{x})\}_{x = F_j(\mathbf{p}_l)}$$

and hence

$$(f,v) - (f,v)_h = \sum_{j=1}^{S} E_0(fvJ_j). \quad (5.27)$$

Again if $f \in \mathscr{H}_2^{(k)}(T_0)$ for some k, it is possible to apply the quadrature error bounds to this form of product provided that vJ_j is a polynomial in **p**. Once we have obtained bounds on the perturbations (5.26) and (5.27) these bounds can be used in (5.15) or (5.17) to estimate the convergence of the approximation.

If for each element the transformation from the standard element is linear, then the Jacobians J_j are all constant and if $W_h(\mathbf{x})$ is a polynomial so are $\partial W_h(\mathbf{p})/\partial x$, $\partial W_h(\mathbf{p})/\partial y$ and $W_h(\mathbf{p})$. It can be shown that in each element the functions $\partial W_h(\mathbf{p})/\partial x\, J(\mathbf{p})$, $\partial W_h(\mathbf{p})/\partial y\, J(\mathbf{p})$ and $W_h(\mathbf{p})J(\mathbf{p})$ are all polynomials when $W_h(\mathbf{x})$ is a finite element trial function of various kinds (Ciarlet and Raviart, 1972c).

Exercise 19 Verify that, for the quadratic isoparametric element given in Exercise 9, it follows that $J(\partial p/\partial x)$, $J(\partial p/\partial y)$, $J(\partial q/\partial x)$ and $J(\partial q/\partial y)$ are all *linear* functions of p and q. Hence verify that for any piecewise quadratic trial function W_h, it follows that in each triangle the functions $J(\mathbf{p})\, \partial W_h(\mathbf{p})/\partial x$ and $J(\mathbf{p})\, \partial W_h(\mathbf{p})/\partial y$ are *quadratic* polynomials in **p**; also verify that $J(\mathbf{p})\, W_h(\mathbf{p})$ is a fourth-order polynomial.

Exercise 20 Prove that if triangular isoparametric elements of degree k are used in each triangle; J, $J(\partial W_h/\partial x)$ and $J(\partial W_h/\partial y)$ are all polynomials of degree $2(k-1)$ for any trial function W_h. Prove that, in general, $JW_h \in P_{3k-2}$ in each triangle.

Given the assumption on the polynomial nature of $J(\mathbf{p})\partial W_h/\partial x$, $J(\mathbf{p})\partial W_h/\partial y$ and $J(\mathbf{p})W_h$ in each element, it is possible to derive bounds on the perturbations (5.26) and (5.27).

Theorem 5.5 Assume that for any trial function $W_h \in K_h$ it follows that in each element, $J(\mathbf{p})\partial W_h/\partial x$ and $J(\mathbf{p})\partial W_h/\partial y$ are polynomials of degree at most r_1; that $J(\mathbf{p})W_h$ is a polynomial of degree at most r_0 and that the regularity hypothesis is valid for all $s \leqslant \max\{r_1, r_0\}$. Then if the quadrature scheme (on the standard triangle) is exact for all polynomials of degree at most $r_1 + s$, it follows that the perturbation (5.26) in the bilinear form, is bounded as

$$\frac{|(a - a_h)(V_h, W_h)|}{\|W_h\|_{1,R}} \leqslant Ch^{s+1}\, \|V_h\|_{s+2,R}, \tag{5.28}$$

where $V_h, W_h \in K_h$ are any trial functions. Similarly, if the quadrature is exact for all polynomials of degree at most $r_0 + s - 1$, then the perturbation (5.27) in the right-hand side, is bounded as

$$\frac{|(f, W_h) - (f, W_h)_h|}{\|W_h\|_{1,R}} \leqslant Ch^{s+1}\, \|f\|_{s+1,R}, \tag{5.29}$$

for any $W_h \in K_h$ provided that $f \in \mathscr{H}_2^{(s+1)}(R)$.

If for example, we use *Langrange or Hermite elements of degree k*, the transformation from T to T_0 is linear and in each element $J_0(\partial W_h/\partial x) \in P_{k-1}$, $J_0(\partial W_h/\partial y) \in P_{k-1}$ and $J_0 W_h \in P_k$. Thus $r_1 = k - 1$, $r_0 = k$ and hence if we use a quadrature that is *exact for all polynomials of degree at most $2k - 2$*, it follows that $s = k - 1$ in (5.28) and (5.29). It then follows from (5.17) that

$$\| u - U_h \|_{1,R} \leqslant C \| u - \tilde{u} \|_{1,R} + \sup_{W_h \in K_h} \left\{ \frac{| (a - a_h)(\tilde{u}, W_h) |}{\| W_h \|_{1,R}} \right\}$$
$$+ \sup_{W_h \in K_h} \left\{ \frac{| (f, W_h) - (f, W_h)_h |}{\| W_h \|_{1,R}} \right\},$$

where $\tilde{u}$ interpolates u. Thus as it is shown in Section 5.3 that one consequence of Theorem 5.4 is

$$\| u - \tilde{u} \|_{1,R} \leqslant C h^k | u |_{k+1,R},$$

we have that the approximation is *optimal* and

$$\| u - U_h \|_{1,R} = O(h^k),$$

since the perturbation terms (5.28) and (5.29) are also $O(h^k)$.

The theorem not only shows the degree of quadrature rule that is necessary to give an optimal approximation; it also shows that the minimum degree necessary to ensure convergence as h tends to zero is $\min\{r_1, r_0 - 1\}$. In the above example this would be $k - 1$.

Proof of Theorem 5.5.   The error in the bilinear form is made up of terms such as

$$E_0 \left(d_1 \frac{\partial V_h}{\partial x} \frac{\partial W_h}{\partial x} J \right) = E_0(vw),$$

where $v(p) = d_1(\partial V_h/\partial x)$ and $w(p) = J(\partial W_h/\partial x)$. If $w \in P_{r_1}$ and the quadrature is exact for polynomials of degree $r_1 + s$, it follows from the Bramble–Hilbert lemma (Exercise 7) that

$$| E_0(vw) | \leqslant C \| w \|_{\mathscr{L}_2(T_0)} | v |_{s+1,T_0}.$$

Assuming that d_1 is sufficiently differentiable, it follows from (5.7a) and Exercise 5 that

$$| E_0(vw) | \leqslant C h^{s+1} \left\| \frac{\partial W_h}{\partial x} \right\|_{\mathscr{L}_2(T)} \left\| \frac{\partial V_h}{\partial x} \right\|_{s+1,T}$$
$$\leqslant C h^{s+1} \| W_h \|_{1,T} \| V_h \|_{s+2,T}.$$

By summing over all elements and dividing by $\| W_h \|_{1,R}$ we obtain (5.28).

In order to obtain (5.29) we follow the proof of Ciarlet and Raviart (1972c) and introduce a projection Π_0 onto the space P_0 of constant functions on T_0; thus for any $u \in \mathscr{L}_2(T_0)$ it follows that

$$\iint_{T_0} (u - \Pi_0 u)\mathrm{d}p = 0.$$

A typical term in the error in the right-hand side of the Galerkin equations can be written as

$$E_0(fW_h J) = E_0(fJ[I - \Pi_0]W_h) + E_0(fJ[\Pi_0 W_h]).$$

Since $J[I - \Pi_0]W_h \in P_{r_0}$, it follows from the Bramble—Hilbert lemma (Exercise 7) that

$$|E_0(fJ[I - \Pi_0]W_h)| \leqslant C \| J[I - \Pi_0]W_h \|_{\mathscr{L}_2(T_0)} |f|_{s, T_0}.$$

But Π_0 is a projection operator and so it is possible to apply Lemma 5.5 to obtain

$$\| [I - \Pi_0]W_h \|_{\mathscr{L}_2(T_0)} \leqslant C |W_h|_{1, T_0}.$$

It follows from (5.7b) therefore, that

$$|E_0(fJ[I - \Pi_0]W_h)| \leqslant C \left\{ \sup_{p \in T_0} J(p) \right\} |f|_{s, T_0} |W_h|_{1, T_0}$$

$$\leqslant Ch^{s+1} \| f \|_{s, T} \| W_h \|_{1, T}. \tag{5.30}$$

Similarly as $J(p) \in P_{r_0}$ and $\Pi_0 W_h$ is a constant it follows that

$$|E_0(fJ[\Pi_0 W_h])| \leqslant C \| J[\Pi_0 W_h] \|_{\mathscr{L}_2(T_0)} |f|_{s+1, T_0}$$

$$\leqslant C \left\{ \sup_{p \in T_0} J(p) \right\} \| W_h \|_{\mathscr{L}_2(T_0)} |f|_{s+1, T_0}$$

$$\leqslant Ch^{s+1} \| f \|_{s+1, T} \| W_h \|_{\mathscr{L}_2(T)}. \tag{5.31}$$

Combining (5.30) with (5.31) summing over all elements and dividing by $\| W_h \|_{1, R}$ leads to the desired result.

Exercise 21   By considering errors of the form

$$E_0 \left(\frac{\partial^2 V_h}{\partial x^2} \frac{\partial^2 W_h}{\partial x^2} J \right),$$

show that if the quadrature is exact for polynomials of degree $r_1 + s$ and the transformation from an arbitrary element onto the standard element is *linear* then the perturbation $(a - a_h)$ for fourth-order problems, can be bounded as

$$\frac{(a - a_h)(V_h, W_h)}{\| W_h \|_{2, R}} \leqslant Ch^{s+1} \| V_h \|_{s+3, R},$$

provided the trial functions are conforming and polynomials of degree at most $r_1 + 2$ in each element.

Results similar to Theorem 5.5 have been obtained by Fix (1972) in the study of the effect of quadrature formulae with both Lagrangian and Hermitian finite element approximations for a polygonal region. Quadrature schemes have also been studied by Herbold and Varga (1972), but only for rectangular regions and assuming that the bilinear form a_h was integrated exactly.

Ciarlet and Raviart (1972c) have applied the results of Theorem 5.5 to isoparametric approximations defined in terms of both triangles and quadrilaterals. They also show how quadrature schemes can be chosen such that the bilinear form a_h is K_h-elliptic and hence it is possible to justify the use of the error estimate (5.17).

These results, however, lead to useful error bounds only when isoparametric elements are used with, at most, *one curved side*. Even in such cases the results are subject to the conditions outlined in Section 5.3. Most of the quadrature error bounds generalize to $\mathbb{R}^m$ $(m > 2)$ for differential equations of the form

$$\sum_{i,\,j=1}^{m} \frac{\partial}{\partial x_i}\left(d_{ij}\frac{\partial u}{\partial x_j}\right) + f(\mathbf{x}) = 0,$$

where $\mathbf{x} = (x_1, \ldots, x_m)$.

Exercise 22 Using the results of Exercise 20, verify that an isoparametric approximation of degree k is optimal if a quadrature rule of degree $4(k-1)$ is used.

(B) Interpolated boundary conditions

We now assume that the approximating subspace K_h contains functions that do not vanish on the boundary but that the integrals are evaluated exactly. Thus it is possible to use the error bound (5.15) in which the perturbation term is

$$\sup_{W_h \in K_h}\left\{\frac{|\,(f,W_h) - a(u,W_h)\,|}{\|W_h\|}\right\}. \tag{5.32}$$

We consider initially the approximate solution of

$$a(u,v) = (f,v),$$

where

$$a(u,v) = \iint_R \left\{\left(\frac{\partial u}{\partial x}\right)\left(\frac{\partial v}{\partial x}\right) + \left(\frac{\partial u}{\partial y}\right)\left(\frac{\partial v}{\partial y}\right)\right\}dx\,dy,$$

corresponding to the differential equation

$$\frac{\partial^2 u}{\partial x^2} + \frac{\partial^2 u}{\partial y^2} + f(x,y) = 0 \quad ((x,y)\in R),$$

subject to

$$u = 0 \quad ((x,y) \in \partial R).$$

The results of this section can be applied with inhomogeneous boundary conditions if, as in Section 5.1, a suitable continuation of the boundary data is available. An alternative approach might be to use Theorem 5.3 to estimate the errors.

An analysis of methods that do not satisfy the boundary conditions exactly, nearly always involves boundary integrals. This is also true of the penalty methods described in Section 5.4(D). It is possible to define a Sobolev space $\mathcal{H}_2^{(k-1)}(\partial R)$ analagous to the space $\mathcal{H}_2^{(k-1)}(R)$ defined in Section 5.1. It follows from such a definition that, for example,

$$\left| \int_{\partial R} \left(\frac{\partial u}{\partial n} \right) W_h \, d\sigma \right| \leqslant \left\| \frac{\partial u}{\partial n} \right\|_{k-1,\partial R} \| W_h \|_{1-k,\partial R}$$

then by the trace theorem, it follows that

$$\left| \frac{\partial u}{\partial n} \right|_{k-1,\partial R} \leqslant C \| u \|_{k,R}.$$

It then follows from Green's theorem, that

$$| (f,W_h) - a(u,W_h) | = \left| \int_{\partial R} \left(\frac{\partial u}{\partial n} \right) W_h \, d\sigma \right|$$

and hence by combining these two expressions, we obtain bounds of the form

$$| (f,W_h) - a(u,W_h) | \leqslant C \| u \|_{k+1,R} \| W_h \|_{-k,\partial R}.$$

To be of any use in second-order problems, (5.32) must involve $\| W_h \|_{1,R}$ and not $\| W_h \|_{-k,\partial R}$. For this reason Scott (1975) has derived bounds of the form

$$\sup_{W_h \in K_h} \left\{ \frac{\| W_h \|_{1-k,\partial R}}{\| W_h \|_{1,R}} \right\} \leqslant Ch^{k+\frac{1}{2}}, \tag{5.33}$$

for trial functions W_h that are *nearly zero* on the boundary ∂R. Berger, Scott and Strang (1972) proved (5.33) for the particular case $k = 1$, but as they did not make the best possible choice of trial functions they were unable to generalize the result. From the definitions of a dual space, it follows that (5.33) is equivalent to

$$\sup_{W_h \in K_h} \left\{ \sup_{g \in \mathcal{H}_2^{(k-1)}(\partial R)} \left\{ \frac{\left| \int_{\partial R} g W_h \, d\sigma \right|}{\| g \|_{k-1,\partial R} \| W_h \|_{1,R}} \right\} \right\} \leqslant Ch^{k+\frac{1}{2}},$$

that is, for all $g \in \mathcal{H}_2^{(k-1)}(\partial R)$ and $W_h \in K_h$

$$\left| \int_{\partial R} gW_h \, d\sigma \right| \leqslant Ch^{k+\frac{1}{2}} \| g \|_{k-1, \partial R} \| W_h \|_{1, R}. \tag{5.34}$$

Lagrangian elements

We now verify (5.34) for the particular type of interpolated boundary conditions considered by Scott; in general we follow his analysis. The region R is partitioned into triangular elements that have straight sides in the interior of R, but which may have a single curved side if the element is adjacent to the boundary. In this section it is assumed that the (curved) boundary can be parametrized in any boundary element T_j as

$$\partial R_j = \{(x_j(\theta), y_j(\theta)) : \theta \in [0, \Theta_j]\}.$$

To simplify the algebra, it is assumed that in a typical boundary element T, the coordinates x and y are such that $\theta = x$ (see Figure 25). Then the arc length $\sigma(x)$ can be written as

$$\sigma(x) = \int_0^x \{1 + y'^2\}^{1/2},$$

and where

$$d\sigma = \frac{d\sigma(x)}{dx} dx.$$

The data points on the curved side of an element are defined by means of the nodes of $(k + 1)$-point *Lobatto quadrature formulae* on the interval $[0,1]$; these nodes are denoted by $\xi_i^{[k]}$ $(i = 1, \ldots, k + 1)$. Davis and Rabinowitz (1967) for example, give a discussion of such quadrature formulae. Thus in a typical element the data points are $(\Theta\xi_i^{[k]}, y(\Theta\xi_i^{[k]}))\dagger$ $(i = 1, \ldots, k + 1)$.

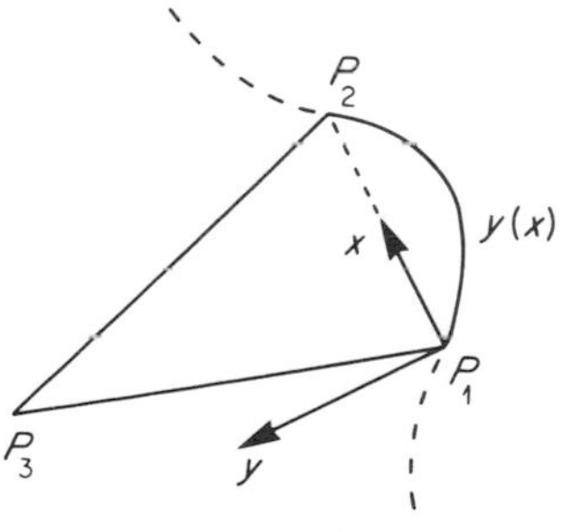

Figure 25

†Scott has shown that this condition can be relaxed somewhat without affecting the final result.

Table 3 Table of Lobatto quadrature points [0,1]	
$k = 2$	0, 1
$k = 3$	0, ½, 1
$k = 4$	$0, \frac{1}{2} \pm \dfrac{1}{2\sqrt{5}}, 1$
$k = 5$	$0, \frac{1}{2} \pm \frac{1}{2} \dfrac{3}{\sqrt{7}}, \frac{1}{2}, 1$

It is possible to rewrite the boundary integral in (5.34) as

$$\int_{\partial R} gW_h \, d\sigma = \sum_j \int_{\partial R_j} gW_h \, d\sigma. \tag{5.35}$$

A typical term in the summation can be written as

$$\int_0^{\Theta} z(x) w(x) \, dx, \tag{5.36}$$

where

$$z(x) = g(x, y(x)) \frac{d\sigma}{dx}$$

and

$$w(x) = W_h(x, y(x)).$$

We now introduce a polynomial $\tilde{z}(x) \in P_{k-2}$ that interpolates $z(x)$ on the interval $[0,\Theta]$. It follows from Theorem 5.4 (with $S = 1$ and $R = (0,\Theta) = I$) that

$$\| z - \tilde{z} \|_{r,I} \leqslant C\Theta^{k-r-1} \| z \|_{k-1,I} \quad (r \leqslant k-2), \tag{5.37}$$

provided that $z(x)$ and hence $g(x,y)$ and $\sigma(x)$ are sufficiently differentiable. Thus it is possible to write (5.36) as

$$\int_0^{\Theta} zw \, dx = \int_0^{\Theta} (z - \tilde{z})w \, dx + \int_0^{\Theta} \tilde{z}w \, dx. \tag{5.38}$$

Since the transformation from the interval $\bar{I} = [0,\Theta]$, onto the standard interval $[0,1]$, can obviously be linear, it follows from the Bramble–Hilbert lemma (Exercise 23) that

$$\left| \int (z - \tilde{z})w \, dx \right| \leqslant C\Theta^{k-1} \| w \|_{\mathscr{L}_2(I)} \, |z|_{k-1,I}.$$

But $W_h(x,y)$ vanishes at the $(k + 1)$ boundary data points and hence

$w(x)$ vanishes at $(k + 1)$ points in the interval I, hence from Rollé's theorem

$$| w(x) | \leqslant C\Theta^{k+1} \left\{ \max_{[0,1]} \left| \frac{d^k w}{dx^k} \right| \right\}.$$

Therefore, as Θ cannot exceed the diameter of the element T, this leads to

$$| w(x) | \leqslant Ch^{k+1} \| W_h \|_{(k)\overline{T}}. \tag{5.39}$$

It can be shown (Exercise 24) that,

$$h^k \| W_h \|_{(k)\overline{T}} \leqslant C \| W_h \|_{1,T}. \tag{5.40}$$

Combining the last two inequalities and integrating, we have that

$$\| w \|_{\mathscr{L}_2(I)} \leqslant Ch^{3/2} \| W_h \|_{1,T}$$

and so

$$\left| \int (z - \tilde{z})w \, dx \right| \leqslant Ch^{k+\frac{1}{2}} \| g \|_{k-1,\partial R_T} \| W_h \|_{1,T}, \tag{5.41}$$

where ∂R_T denotes the curved part of the boundary of the element T.

We now consider the second term in (5.38). Since $w(x)$ vanishes at the Lobatto quadrature points, it follows that an integral involving $w(x)$ in the integrand is approximated by zero using Lobatto quadrature. A bound on the quadrature error is therefore a bound on the value of the analytic integral. Thus, as $(k + 1)$-point Lobatto quadrature is exact for polynomials of degree at most $2k - 1$, it follows from the Bramble—Hilbert lemma (Exercise 6) that

$$\left| \int_0^{\Theta} \tilde{z}(x)w(x)dx \right| \leqslant Ch^{2k} \, | \tilde{z}w |_{2k,I};$$

but $\tilde{z} \in P_{k-2}$ and $w(x) = W_h(x, y(x))$, where $W_h \in P_k$ in T, hence

$$| \tilde{z}w |_{2k,I} \leqslant C \| \tilde{z} \|_{k-2,I} \| W_h \|_{k,I}.$$

With (5.37), this leads to

$$\| \tilde{z} \|_{k-2,I} \leqslant \| z \|_{k-2,I} + \| \tilde{z} - z \|_{k-2,I}$$
$$\leqslant C \| z \|_{k-1,I}$$
$$\leqslant C \| g \|_{k-1,\partial R_T}.$$

Since the transformation from T onto the standard element T_0 is linear it follows from the regularity hypothesis that

$$\| W_h \|_{k,T} \leqslant Ch^{1-k} \| W_h \|_{1,T}.$$

Combining these four inequalities, we have that

$$\left| \int_0^\Theta \tilde{z}w \ dx \right| \leqslant Ch^{k+1} \ \|g\|_{k-1, \partial R_T} \ \|W_h\|_{1, T}. \tag{5.42}$$

Thus summing terms such as (5.42) and (5.41) over all the boundary elements, we have that the perturbation error is bounded as

$$\left| \int_{\partial R} gW_h \ d\sigma \right| \leqslant Ch^{k+\frac{1}{2}} \ \|g\|_{k-1, \partial R} \ \|W_h\|_{1, R} \tag{5.43}$$

and hence

$$\left\{ \frac{|(f, W_h) - a(u, W_h)|}{\|W_h\|_{1, R}} \right\} \leqslant Ch^{k+\frac{1}{2}} \ \|u\|_{k+1, R}. \tag{5.44}$$

Theorem 5.4 shows that Lagrange interpolating polynomials of degree k lead to interpolation error bounds of the form

$$\|u - \tilde{u}\|_{1, R} \leqslant Ch^k \ |u|_{k+1, R}$$

and hence the Galerkin error bounds are of the form

$$\|u - U_h\|_{1, R} \leqslant C\{h^k \ |u|_{k+1, R} + h^{k+\frac{1}{2}} \ \|u\|_{k+1, R}\}. \tag{5.45}$$

So a finite element approximation with such interpolated boundary conditions is *optimal* since the perturbation error is of a higher order (in h) than the approximation error. Scott (1975) and Chernuka, Cowper, Lindberg and Olson (1972) have devised quadrature rules for triangular elements with curved boundaries that preserve the order for piecewise quadratic approximations. Piecewise quadratics have also been studied by Berger (1973) to derive a bound on the error in terms of the $\mathscr{L}_2(R)$ norm; he has also verified the orders numerically (1972). As an alternative to interpolating the boundary data at a finite point set, the approximation could be made to match along the entire boundary if blending function interpolants are used (Gordon and Wixom, 1974). Blending functions are discussed in some detail in Section 7.3.

Exercise 23   By using an analysis similar to that of Exercise 6 (or Lemma 5.5), prove that it follows from the Bramble—Hilbert lemma that

$$\left| \int_0^\Theta (z - \tilde{z})w \ dx \right| \leqslant C\Theta^{k-1} \ \|w\|_{\mathscr{L}_2(I)} \ |z|_{k-1, I},$$

where $\tilde{z} \in P_{k-2}$ interpolates z on $I = (0, \Theta)$ and $w \in \mathscr{L}_2(I)$. (Hint. Consider the functional $F \in \mathscr{L}(\mathscr{H}_2^{(k-1)}(I) \times \mathscr{L}_2(I); \mathbb{R})$, where

$$F(z, w) = \int_0^\Theta (z - \tilde{z})w \ dx.)$$

Exercise 24   With $k = 2$ and $k = 3$, verify that for Lagrange interpolating polynomials of degree k in triangles with straight sides, any
trial function W satisfies

$$\| W \|_{(k)\bar{T}} \leqslant Ch^{-k} \| W \|_{1,T}.$$

Exercise 25   Verify that the bound on the perturbation error, given by
(5.44), is also valid for the particular form of Hermitian-type cubic
elements devised by Scott (1975).

(C) Boundary approximation

Probably the first error bounds for finite element methods over
perturbed regions, were obtained by Russian mathematicians (for
example Oganesyan, 1966). They derived bounds for piecewise linear
approximations based on triangular meshes and they considered only
the approximate solution of second-order problems subject to the
boundary condition

$$\frac{\partial u}{\partial n} + \beta u = 0 \quad (\beta \geqslant 0)$$

on a curved boundary. They showed that

$$\| u - U_h \|_{1,R_h} \leqslant Ch \| u \|_{2,R_h},$$

but the proofs are rather technical and are beyond the scope of this
book. Others (Oganesyan and Rukhovets, 1969) have also derived
error bounds for this type of problem in terms of the $\mathscr{L}_2(R)$ norm.
More recently, it has been shown (Strang and Berger, 1971; Thomée,
1973) that if R_h is a polygon inscribed in $R \subset \mathbb{R}^2$ ($\mathbb{R}^m, m \geqslant 2$
according to Strang and Fix, 1973, p. 196) then for the model problem
of

$$\frac{\partial^2 u}{\partial x^2} + \frac{\partial^2 u}{\partial y^2} + f(x,y) = 0 \quad ((x,y) \in R) \tag{5.46}$$

subject to $u = 0$ on ∂R, it follows that

$$\| u - u_h \|_{1,R_h} = O(h^{3/2}),$$

where u_h is the solution of the perturbed problem consisting of

$$\frac{\partial^2 u_h}{\partial x^2} + \frac{\partial^2 u_h}{\partial y^2} + f(x,y) = 0 \quad ((x,y) \in R_h) \tag{5.47}$$

subject to $u_h = 0$ on ∂R_h.

Thus if (5.46) is solved approximately, by partitioning the polygonal
region R_h and then computing a finite element solution of (5.47), it

follows from Section 5.3 that

$$\| u - U_h \|_{1, R_h} = O(h)$$

for piecewise linear approximations based on a triangular partition. Similarly it follows that if approximating functions include all polynomials of degree 2, or higher, then

$$\| u - U_h \|_{1, R_h} = O(h^{3/2}).$$

This order of approximation may be significantly lower than that expected from Section 5.3 and it arises from a poor approximation near the boundary; sometimes referred to as a *boundary layer effect*. Maximum principles can be used to verify that the perturbations are smaller in the interior of R, where sharper bounds are available. Some of these results can be extended to the situation when $R_h \not\subset R$. The convergence properties of the finite element approximations away from the boundary have also been studied by Nitsche and Schatz (1974) and Bramble and Thomée (1974).

Berger, Scott and Strang (1972) show that, in general, if the region R is approximated by R_h — which is not necessarily a polygon — such that the maximum distance between the two boundaries ∂R and ∂R_h is $O(h^{k+1})$ then the perturbation term in (5.16) is $O(h^{k+\frac{1}{2}})$. As a piecewise k-degree polynomial approximation of the boundary — by interpolation say — could satisfy this condition; it follows that if the same degree of polynomials are used in both the function approximation and the boundary interpolation, then the finite element error is still $O(h^k)$ overall in terms of the $\mathcal{H}_2^{(1)}(R_h)$ norm. An analagous result is true for isoparametric approximations since Ciarlet and Raviart (1972c) have shown that the conclusions of Theorem 5.5, subject to minor modifications, are valid when the region is perturbed. A similar result has been obtained by Zlámal (1973) and (1974). The Neumann problem has been considered by a few authors such as Strang and Fix (1973) and Babuška (1971).

(D) Penalty methods

This category includes all methods that incorporate non-homogeneous Dirichlet boundary conditions in the form of a boundary integral that is added to the functional, rather than as a condition to be imposed on the approximating functions. Such methods can be based on the method of least squares or the Ritz method, or a combination of both. The most frequently used formulation is based on the method of least squares for which the errors are no longer derived naturally in terms of Sobolev norms and invariably involve the *trace theorem* to deal with the boundary integrals (for example Varga, 1971, Chapter 6).

If we assume that the interpolation error bound given by Theorem

5.4 is valid for some $k > 0$, that is,

$$\| u - \tilde{u} \|_{r, R} \leqslant Ch^{k+1-r} \| u \|_{k+1, R},$$

then it is possible to derive a straightforward bound on the error — but not in terms of Sobolev norms.

Theorem 5.6   If the finite element approximation satisfies

$$(AU_h - f, AV_h) = h^{-3} \langle U_h - g, V_h \rangle \quad \text{(for all } V_h \in K_h\text{)}, \qquad (5.48)$$

where

$$\langle \phi, \psi \rangle = \int_{\partial R} \phi \psi \, d\sigma,$$

then if Theorem 5.4 is valid for some $k > 0$, it follows that

$$\| AU_h - Au \|_{\mathscr{L}_2(R)} + h^{3/2} \| U_h - u \|_{\mathscr{L}_2(\partial R)} \leqslant Ch^{k-1} \| u \|_{k+1, R}.$$

Proof   Since

$$\| Au - A\tilde{u} \|_{\mathscr{L}_2(R)} \leqslant C \| u - \tilde{u} \|_{2, R}$$

for any $(u - \tilde{u}) \in \mathscr{H}_2^{(2)}(R)$ and

$$\| u - \tilde{u} \|_{\mathscr{L}_2(\partial R)} \leqslant C \{ h^{-1/2} \| u - \tilde{u} \|_{\mathscr{L}_2(R)} + h^{1/2} \| u - \tilde{u} \|_{1, R} \}$$

(Agmon, 1965); it follows that

$$\| Au - A\tilde{u} \|_{\mathscr{L}_2(R)} + h^{3/2} \| u - \tilde{u} \|_{\mathscr{L}_2(R)}$$
$$\leqslant C \{ \| u - \tilde{u} \|_{2, R} + h^{-1} \| u - \tilde{u} \|_{1, R} + h^{-2} \| u - \tilde{u} \|_{\mathscr{L}_2(R)} \}. \qquad (5.49)$$

Since the least squares method given by (5.48), is a *projection method* in the sense that

$$\| u - U_h \|_{[1]} = \inf_{\tilde{u} \in K_h} \| u - \tilde{u} \|_{[1]},$$

where the norm is defined by

$$\| \phi \|_{[1]}^2 = \| A\phi \|_{\mathscr{L}_2(R)}^2 + h^{-3} \| \phi \|_{\mathscr{L}_2(\partial R)}^2.$$

The result follows immediately from (5.49) and Theorem 5.4 with $r = 0, 1$ and 2.

It is also possible to prove that

$$\| u - U_h \|_{\mathscr{L}_2(R)} \leqslant Ch^{k+1} \| u \|_{k+1, R},$$

but the proof is beyond the scope of this book (Baker, 1973; or Bramble and Schatz, 1970). A numerical example of an application of this particular form is given in Section 7.4(A); others can be found in Serbin (1975).

Other authors have suggested alternative projection methods for

138

solving (5.46); they have based their methods on such norms as

$$\| \phi \|_{[2]}^2 = a(\phi,\phi) + h^{-1} \| \phi \|_{\mathscr{L}_2(\partial R)}^2$$

(Bramble, Dupont and Thomée, 1972) and

$$\| \phi \|_{[3]}^2 = -a(\phi,\phi) - 2(A\phi,\phi) + h^2 \| A\phi \|_{\mathscr{L}_2(R)}^2$$

$$+ \gamma \left\{ h^{-1} \| \phi \|_{\mathscr{L}_2(\partial R)}^2 + h \left\| \frac{\partial \phi}{\partial s} \right\|_{\mathscr{L}_2(\partial R)}^2 \right\} \qquad (\gamma > 0)$$

(Bramble and Nitsche, 1973).

Such methods have been extended to problems of higher degree and also to problems in more than two dimensions. Methods based on stationary points of functionals that are not positive definite have also been suggested (for example Thomée, 1973). Penalty methods have also been studied by Aubin (1972) p. 17.

(E) Non-conforming elements

We define

$$a_h(u,v) = \sum_{j=1}^S \iint_{T_j} \left\{ \frac{\partial u}{\partial x} \frac{\partial v}{\partial x} + \frac{\partial u}{\partial y} \frac{\partial v}{\partial y} \right\} dx\,dy,$$

which does not equal

$$a(u,v) = \iint_R \left\{ \frac{\partial u}{\partial x} \frac{\partial v}{\partial x} + \frac{\partial u}{\partial y} \frac{\partial v}{\partial y} \right\} dx\,dy,$$

if v has a jump discontinuity across the inter-element boundaries; it is the difference between the two forms a and a_h that is crucial in the study of non-conforming elements (Section 7.2); in addition we define a semi-norm corresponding to a_h by

$$| u |_h = [a_h(u,u)]^{1/2}$$

and also a norm

$$\| u \|_h = [| u |_h^2 + \| u \|_{\mathscr{L}_2(R)}^2]^{1/2}.$$

As an example of the results that can be obtained for non-conforming elements we consider the space K_h of piecewise linear elements that are matched at the *side mid-points* of a triangular mesh, where again further details can be found in Section 7.2. It follows with such elements, that the error in interpolating linear functions is zero, so by analogy with Section 5.3

$$\inf_{V_h \in K_h} \| u - V_h \|_h \leqslant Ch \, | u |_{2,R}$$

for any $u \in \mathscr{H}_2^{(2)}(R)$. Thus if the non-conforming approximation is to be optimal, the perturbation term

$$\sup_{W_h \in K_h} \frac{|(f, W_h) - a_h(u, W_h)|}{\| W_h \|_h}$$

must be at least $O(h)$.

By applying Green's theorem in each element it follows that

$$(f, W_h) - a_h(u, W_h) = \sum_{j=1}^{S} \int_{\partial T_j} \frac{\partial u}{\partial n} W_h \, \partial \sigma$$

$$= \sum_{l=1}^{Q} \int_{E_l} \left\{ \left(\frac{\partial u}{\partial n} W_h \right)^{[1]} + \left(\frac{\partial u}{\partial n} W_h \right)^{[2]} \right\} d\sigma,$$

where E_l denotes a particular edge of the mesh and the two terms in the final integral, numbered [1] and [2] respectively, are the limit values corresponding to the elements on either side of the interface E_l. By applying the Bramble—Hilbert lemma to functionals of the form

$$F(u, W_h) = \int_{E_l} \left\{ \left(\frac{\partial u}{\partial n} W_h \right)^{[1]} + \left(\frac{\partial u}{\partial n} W_h \right)^{[2]} \right\} d\sigma \qquad (5.50)$$

(Exercise 27), it is possible to show (Crouzeix and Raviart, 1973) that if

$$\int_{E_l} \{ W_h^{[1]} - W_h^{[2]} \} d\sigma = 0 \qquad (5.51)$$

for all $W_h \in K_h$, then

$$|(f, W_h) - a_h(u, W_h)| \leqslant Ch \, |u|_{1, R} \, \| W_h \|_h. \qquad (5.52)$$

Exercise 26 Verify that (5.51) is satisfied for the piecewise linear non-conforming elements described above.

Note that it is shown in Section 7.2 that (5.51) is related to the *patch test* for non-conforming elements for second-order problems. Ciarlet (1973) has produced similar results for plate bending elements. A comprehensive analysis of non-conforming methods for plate-bending problems has been provided by Lascaux and Lesaint (1975).

Exercise 27 Assuming (5.51), apply the Bramble—Hilbert lemma to (5.50) and hence prove (5.52).

5.5 SUMMARY

This section contains a brief summary of some of the important results of Sections 5.2—5.4.†

---

†Although this section can be read before the preceding sections of the chapter, the results given are expressed using the notation described in Section 5.1.

140

The first result depends on the interpolation properties of the basis functions used in the finite element approximation — the completeness hypothesis. In general, if the error in interpolating a polynomial of degree k is zero, and if the solution u, of a second-order problem in a two-dimensional region R, is sufficiently smooth so that $u \in \mathscr{H}_2^{(k+1)}(R)$, then the Galerkin approximation $U \in K_N$ is such that

$$\| u - U \|_{1,R} \leqslant C \| u - \tilde{u} \|_{1,R} \leqslant Ch^k \| u \|_{k+1,R}, \tag{5.53}$$

where h is the diameter of the largest element and $\tilde{u} \in K_N$ interpolates u. In this context, the form of the interpolating function is governed by the type of finite element approximation being used; that is, interpolation implies that all the nodal parameters are specified exactly.

When the solution u does not have the degree of smoothness demanded by (5.53), the exponent of h in the error estimate is reduced such that if $u \in \mathscr{H}_2^{(k^*+1)}(R)$ $(k^* \leqslant k)$ then

$$\| u - U \|_{1,R} \leqslant Ch^{k^*} \| u \|_{k^*+1,R}. \tag{5.54}$$

This modified form of error estimate has to be used when, for example, there are singularities in any of the low-order derivatives of the solution inside the region R or on the boundary. The case where there are boundary singularities in $D^i u$ $(|i| \leqslant 1)$ has been discussed extensively elsewhere (Strang and Fix, 1973, Chapter 8 and Wait, 1976) and is considered briefly in Section 7.4(F).

For approximations based on elements in which the basis functions are not specified explicitly as functions of the space variables x and y, but are defined instead by means of transformations onto a standard element in new variables p and q, the exponent of h in (5.53) is defined slightly differently. For such approximations, which include the popular isoparametric elements, k is the degree of polynomials in p and q that can be interpolated exactly.

When using curved isoparametric elements it is important to note the severe restrictions under which estimates (5.53) and (5.54) are satisfied. Even when only one side of a triangular element is replaced by a quadratic curve, the element should be only an $O(h^2)$ perturbation from a straight-sided triangle. If the curved boundary is a cubic polynomial, the conditions on the element are even more severe, details can be found in Section 5.3 on page 121.

When Dirichlet boundary conditions are specified, it is necessary to ensure that the approximate solution satisfies such conditions along *the entire length of the boundary* involved before it is possible to apply (5.53) or (5.54). If alternatively, the boundary conditions are inter-polated at only a finite number of boundary points, the error estimate becomes

$$\| u - U \|_{1,R} \leqslant Ch^k \| u \|_{k+1,R} + \Delta, \tag{5.55}$$

where the size of the perturbation term Δ is governed by the form of the boundary data approximation. The error estimate (5.55) can also be used when the region R is approximated by (say) a polygon, when numerical integration is used, or when non-conforming elements are used in the approximation. All three modifications can be thought of as departures from the classical variational method and so they each fit neatly into this type of analysis. Whatever the reason for the perturbation, convergence is still guaranteed provided the perturbation is $O(h^s)$ $(s > 0)$. The modified form of approximation is said to be *optimal* if the perturbation is $O(h^k)$, that is, if it is of the same order as the interpolation.

(A) Numerical integration

It can be shown that for second-order problems, using straight-sided elements, the perturbation term due to numerical integration is

$$\Delta = O(h^s)$$

provided that the quadrature rule is exact for all polynomials of degree $s + k - 2$, that is, convergence is guaranteed if the rule is of degree† $k - 1$ and optimal if the degree is $2k - 2$. If triangular isoparametric elements with curved sides are used, then subject to the restrictions mentioned earlier, optimal approximations require quadrature rules of degree $4(k - 1)$ whereas convergence is guaranteed if the rule is of degree $3(k - 1)$.

(B) Interpolated boundary conditions

If the boundary conditions are only matched at a finite number of points in each boundary element, then the order of the perturbation term depends on the position of such boundary interpolation points. The approximation is optimal if the boundary data is matched at Lobatto quadrature points; that is, if the boundary of an element can be written parametrically as $(x(\theta), y(\theta))$ $(0 \leqslant \theta \leqslant \Theta)$ then the interpolation points are given by the Lobatto quadrature points on the interval $[0, \Theta]$. Alternatively the quadrature points can be specified in terms of arc length along the boundary.

(C) Boundary approximation for curved boundaries

It can be shown that if the boundary is approximated in each boundary element by a curve whose equation is a polynomial of degree

†A quadrature rule of degree r is a quadrature rule that is exact for all polynomials of degree at most r.

k, the boundary is moved by a distance $O(h^{k+1})$ and then the approximation is optimal. Alternatively if the boundary is approximated by a polygon, the perturbation term is $O(h^{3/2})$ and convergence is still guaranteed.

(D) Non-conforming elements

It can be shown that convergence is guaranteed if the elements pass the patch test (Chapter 7).

Chapter 6

Time-dependent Problems

In Chapter 3 we derive a family of approximate methods of solution for boundary value problems; these depend on obtaining the stationary point of a functional which is also an extremum. In this chapter we extend such methods, where possible, to the solution of initial value problems. However, new problems arise when considering the variational formulation of evolutionary problems. For dissipative systems for example, when the adjoint problem is introduced, the combined functional $I(u,u^*)$ does not have such extremum properties. Even in evolutionary problems for which a true variational principle exists, such as Hamiltonian dynamics, the stationary value is *not* an extremum.

These difficulties have led various authors to suggest that variational formulations should be dispensed with for the solution of time-dependent problems, as they are of little practical significance. We subscribe to this view, for, as we show in Chapter 3, Galerkin's method can be introduced without any reference to variational principles. The sole justification for introducing the adjoint problem is to bring dissipative systems within the scope of a particular mathematical structure. With this goal in view, other authors have suggested so-called restricted variational principles or quasi-variational principles; such principles have little intrinsic value, but simply provide a mathematical justification for the Galerkin method for dissipative systems. All the formulations are equally good in this respect, and are equally lacking in rigour as far as the treatment of initial value problems are concerned.

6.1 HAMILTON'S PRINCIPLE

In Section 2.5 the equations of motion of a continuous dynamical system are derived as necessary conditions for a stationary point of a functional. We now show that if this formulation is used an approximate solution can be obtained as in Chapter 3, by determining a stationary value with respect to an approximating subspace of functions. As the equations of motion of such dynamical systems are of hyperbolic or parabolic type, it follows that the corresponding functional is not positive definite. Hence even for conservative systems the stationary value does *not* give an extremum and it is not possible to derive a best

144

approximation. An example of such a functional corresponding to the equation of motion

$$\frac{\partial^2 u}{\partial t^2} - c^2 \frac{\partial^2 u}{\partial x^2} = 0 \qquad (6.1)$$

for a vibrating string is

$$I(v) = \tfrac{1}{2}\rho \int_{t_0}^{t_1} \int_0^l \left[\left(\frac{\partial v}{\partial t}\right)^2 - c^2 \left(\frac{\partial v}{\partial x}\right)^2 \right] \mathrm{d}x \mathrm{d}t. \qquad (6.2)$$

Boundary conditions

It would appear that in order to obtain an approximate solution of the form

$$U(x,t) = \sum_{i=1}^{N} \alpha_i \varphi_i(x,t)$$

to (6.1) we simply apply the Ritz method of Section 3.1 with the functional $I(v)$ given in (6.2); unfortunately this leads to inconsistencies in the boundary conditions.

The solution of a linear *hyperbolic* differential equation of the form

$$\frac{\partial^2 u(\mathbf{x},t)}{\partial t^2} + Au(\mathbf{x},t) = f(\mathbf{x},t) \quad (t_0 < t \leqslant t_1), \qquad (6.3)$$

for $\mathbf{x} \in R \subset \mathbb{R}^m$ with boundary ∂R, where A is a linear elliptic differential operator similar to those introduced in Chapter 3, is well defined in $R \times [t_0, t_1]$ if the conditions

$$u(\mathbf{x},t) = 0 \qquad ((\mathbf{x},t) \in \partial R \times (t_0, t_1)), \qquad (6.3a)$$

$$u(\mathbf{x},t_0) = u_0(\mathbf{x}) \quad (\mathbf{x} \in R) \qquad (6.3b)$$

and

$$\frac{\partial u(\mathbf{x},t_0)}{\partial t} = v_0(\mathbf{x}) \quad (\mathbf{x} \in R) \qquad (6.3c)$$

hold. However the problem is not well posed if (6.3c) is replaced by a condition at $t = t_1$, such as

$$u(\mathbf{x},t_1) = u_1(\mathbf{x}) \quad (\mathbf{x} \in R). \qquad (6.3d)$$

It is this problem of equation (6.3), subject to (6.3a), (6.3b) and (6.3d) that follows from a stationary value of the corresponding functional

$$I(v:t_0,t_1) = \int_{t_0}^{t_1} \{(v_t,v_t) - a(v,v) + 2(f,v)\} \, \mathrm{d}t, \qquad (6.4)$$

where the inner product notation is as in Chapter 3.

Thus before it is possible to compute the Ritz solution to (6.3) subject to (6.3a), (6.3b) and (6.3c), it is necessary to adapt the method slightly so as to avoid having to introduce the boundary condition (6.3d). We do not encounter this difficulty if we use the Kantorovich (semi-discrete) method to obtain an approximate solution of the form

$$U(\mathbf{x},t) = \sum_{i=1}^{N} \alpha_i(t)\varphi_i(\mathbf{x}),$$

as we replace equation (6.3) by a system of second-order ordinary differential equations in the functions $\alpha_i(t)$ $(i = 1, \ldots, N)$ (Section 3.3). The conditions (6.3b) and (6.3c) are replaced by the equivalent conditions

$$\alpha_i(t_0) = c_i \quad (i = 1, \ldots, N) \tag{6.5a}$$

and

$$\frac{d\alpha_i(t_0)}{dt} = d_i \quad (i = 1, \ldots, N). \tag{6.5b}$$

We leave the discussion of the semi-discrete method until a later section and return briefly to the problem of calculating a Ritz solution of the form

$$U(\mathbf{x},t) = \sum_{i=1}^{N} \alpha_i\varphi_i(\mathbf{x},t). \tag{6.6}$$

If the function $u(\mathbf{x},t)$ is the solution of (6.3), subject to the given conditions then it follows that for any T_0 and T_1 such that $t_0 \leqslant T_0 < T_1 \leqslant t_1$ the function $u(\mathbf{x},t)$ provides a stationary value of the integral $I(v:T_0,T_1)$ where all the admissible functions satisfy the conditions

$$v(\mathbf{x},T_0) = u(\mathbf{x},T_0) \quad (\mathbf{x} \in R)$$

and

$$v(\mathbf{x},T_1) = u(\mathbf{x},T_1) \quad (\mathbf{x} \in R),$$

together with the boundary condition

$$v(\mathbf{x},t) = 0 \quad ((\mathbf{x},t) \in \partial R \times (T_0, T_1)).$$

In particular, if we partition the interval $[t_0, t_1]$ as

$$t_0 = \tau_0 < \tau_1 < \ldots < \tau_K = t_1.$$

then the solution of equation (6.3) provides a stationary value for each of the functionals

$$I_n(v) = \int_{\tau_{n-1}}^{\tau_{n+1}} \{(v_t,v_t) - a(v,v) + 2(f,v)\}\,dt \quad (n = 1, \ldots, K). \tag{6.7}$$

146

Note that the ranges of the functionals are overlapping in time. Using (6.7) in place of (6.4), it is possible to define a step-by-step method of solution based on the Ritz method. If the problem is solved in such a manner, the solution at $t = \tau_{n-1}$ will be available, but the object of computing a stationary value of I would be to find the solution at $t = \tau_{n+1}$ rather than have it available as an *a priori* condition. In practical calculations however we can overcome this objection by ignoring it. We formulate the problem as if the boundary conditions (6.3a) were given together with $u(x,t)$ at $t = \tau_{n-1}, \tau_{n+1}$, then *solve* for the unknown solution at $t = \tau_{n+1}$. This approach is possible provided that we have additional information about the solution at the intermediate step $t = \tau_n$. Thus in order to calculate the solution for $t = \tau_{n+1}$ we must first have the solution at the two previous steps that is $t = \tau_n$ and $t = \tau_{n-1}$.

As an example we describe one way of computing a finite element solution to equation (6.1). We subdivide the region $0 \leqslant x \leqslant l$, $\tau_{n-1} \leqslant t \leqslant \tau_{n+1}$, by means of the partition

$$0 = x_0 < x_1 < \ldots < x_{L+1} = l,$$

where $x_{i+1} - x_i = \Delta x$ $(i = 0,1,2, \ldots , L)$ and where we assume that

$$\tau_{n+1} - \tau_n = \tau_n - \tau_{n-1} = \Delta t.$$

Then we proceed as if to solve the boundary value problem

$$\frac{\partial^2 u}{\partial t^2} - c^2 \frac{\partial^2 u}{\partial x^2} = 0 \quad (0 < x < l; \tau_{n-1} < t < \tau_{n+1}), \tag{6.8}$$

subject to

$$u(x,\tau_{n+1}) = u_{n+1}(x) \quad (0 \leqslant x \leqslant l), \tag{6.9a}$$

$$u(x,\tau_{n-1}) = u_{n-1}(x) \quad (0 \leqslant x \leqslant l) \tag{6.9b}$$

and

$$u(0,t) = u(l,t) = 0 \quad (\tau_{n-1} < t < \tau_{n+1}). \tag{6.9c}$$

In order to obtain an approximate solution of (6.8), subject to (6.9a), (6.9b) and (6.9c), we define in the region $(0 \leqslant x \leqslant l; \tau_{n-1} \leqslant t \leqslant \tau_{n+1})$ bilinear basis functions $\varphi_{ij}(x,t)$ $(i = 1, \ldots , L; j = n-1, n, n+1)$ corresponding to the points $P_i^j = (x_i,\tau_j)$ (cf. Section 3.1) as in Figure 26. The approximate solution is then of the form

$$U(x,t) = \sum_{j=n-1}^{n+1} \sum_{i=1}^{L} \varphi_{ij}(x,t)U_i^j, \tag{6.10}$$

where U_i^j is the approximate solution at the point P_i^j. For solving boundary value problems defined by (6.8)–(6.9c), U_i^{n+1} and U_i^{n-1}

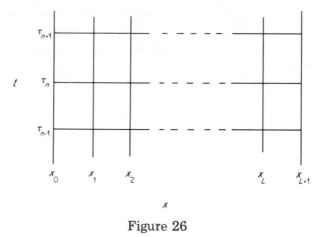

Figure 26

$(i = 1, \ldots, L)$ are determined from (6.9a) and (6.9b) respectively and U_i^n by

$$\frac{\partial}{\partial U_i^n} I_n \left(\sum_{j=n-1}^{n+1} \sum_{k=1}^{L} \varphi_{kj} U_i^j \right) = 0 \quad (i = 1, \ldots, L). \tag{6.11}$$

We are, however attempting to solve an initial value problem, not a boundary value problem. Hence if U_i^{n-1} and U_i^n $(i = 1, \ldots, L)$ are known, (6.11) is used to determine U_i^{n+1}.

Exercise 1 Verify that the step-by-step solution of the equation

$$\frac{\partial^2 u}{\partial t^2} - \frac{\partial^2 u}{\partial x^2} = 0 \quad (0 < x \leqslant l; t_0 < t \leqslant t_1),$$

as described above using bilinear basis functions, leads to the system of difference equations

$$[\delta_t^2 I_x - r^2 \delta_x^2 I_t] U_i^n = 0, \tag{6.12}$$

where δ_t^2 and δ_x^2 are second-order central difference operators, I_x and I_t are the Simpson's rule operators introduced in Section 3.1 and $r = \Delta t/\Delta x$, with the region partitioned such that

$$\tau_{n+1} - \tau_n = \Delta t \quad (n = 0, \ldots, K) \quad \text{and} \quad x_{i+1} - x_i = \Delta x$$

$$(i = 0, \ldots, L).$$

Initial conditions

In the example given above and in step-by-step methods in general derived similarly for (6.3)—(6.3c), it is necessary to approximate the initial conditions in some way to derive the approximate solution at

$t = \tau_0$ and $t = \tau_1$, so that the functional $I_1(v)$ can be used to derive the solution at $t = \tau_2$. If there is no discontinuity between the initial conditions and the boundary conditions, then it is possible to define an approximate solution of the form

$$U(\mathbf{x},t) = u_0(\mathbf{x}) + \sum_{i=1}^{N} \alpha_i \varphi_i(\mathbf{x},t). \qquad (6.13)$$

This procedure is equivalent to transforming the problem (cf. Section 3.2) to one for which the initial condition (6.8b) is replaced by

$$U(\mathbf{x},t_0) = 0 \quad (\mathbf{x} \in R).$$

When there is a discontinuity, that is,

$$u_0(\mathbf{x}) \neq 0 \quad (\mathbf{x} \in \partial R),$$

it is not possible to use an approximation of the form (6.13) and it is necessary to approximate $u_0(\mathbf{x})$ in terms of

$$U(\mathbf{x},t_0) = \sum_{i=1}^{N} \alpha_i \varphi_i(\mathbf{x},t_0),$$

such that either $U(\mathbf{x},t_0)$ interpolates $u_0(\mathbf{x})$ or the error $u_0(\mathbf{x}) - U(\mathbf{x},t_0)$ is minimized in some norm. Similarly, it is necessary to approximate the second initial condition so that

$$\frac{\partial U(\mathbf{x},t_0)}{\partial t} = \sum_{i=1}^{N} \alpha_i \frac{\partial \varphi_i(\mathbf{x},t_0)}{\partial t}$$

is a good approximation to $v_0(\mathbf{x})$.

Thus in the earlier example, it is possible to introduce the initial conditions

$$U_i^0 = u_0(x_i) \quad (i = 1, \ldots, L)$$

and

$$\frac{U_i^1 - U_i^0}{\Delta t} = v_0(x_i) \quad (i = 1, \ldots, L).$$

This use of Hamilton's principle leads to a valid step-by-step method of solution for conservative systems, although the mathematical formulation of such a method is in parts confusing. A similar procedure to describe a step-by-step method of approximation for hyperbolic equations is given by Noble (1973).

6.2 DISSIPATIVE SYSTEMS

As we mentioned at the beginning of this chapter, various authors have attempted to provide a variational formulation for dissipative

problems. Some have attempted to provide a general variational
principle that holds for a large class of such problems (Finlayson and
Scriven, 1967, and references therein) as Hamilton's principle does for
conservative systems. The main drawbacks of such formulations are
that:

(1) The variational principles are derived from the constitutive
     equations rather than the other way round, hence the variational
     formulation involves additional effort, but provides no
     additional information.
(2) The functional involved in the variational principle has no
     physical significance, and the stationary values are never true
     extrema that can be used to derive error bounds.
(3) As indicated in the previous section there are basic discrepancies
     between variational problems and initial value problems that are
     invariably ignored in most so-called variational formulations of
     evolutionary problems.

We have stressed the merits and disadvantages of variational
formulations at this point in the book, in view of the wide variety of
such formulations of *dissipative* or *irreversible* processes in the
literature, and the often contradictory claims made about them. We do
not introduce the finite element solution of evolutionary dissipative
systems using the adjoint formulation instead we provide two exercises
which may be attempted by the interested reader.

Exercise 2   Verify that, with the correct choice of approximating
functions, the functional corresponding to the simple diffusion equation

$$\frac{\partial u}{\partial t} = \frac{\partial^2 u}{\partial x^2} \quad (0 < x < l; t_0 < t \leqslant t_1),$$

subject to

$$u(0,t) = u(l,t) = 0 \quad (t_0 < t \leqslant t_1),$$

together with

$$u(x,t_0) = u_0(x) \quad (0 \leqslant x \leqslant l),$$

is given by

$$I = \int_{t_0}^{t_1} \int_0^l \left\{ \frac{\partial v}{\partial t} v^* + \frac{\partial v}{\partial x} \frac{\partial v^*}{\partial x} \right\} dxdt.$$

Then show that the equation

$$-\frac{\partial u^*}{\partial t} = \frac{\partial^2 u^*}{\partial x^2},$$

is the necessary condition corresponding to the stationary point

$\delta_v I(v,v^*) = 0$, if either (i) u is given for $t = t_0$ and $t = t_1$ or (ii) u is given for $t = t_0$ and $u^* = 0$ for $t = t_1$.

Note. This shows that when calculating a stationary point of I_n, we should assume that $u(x,\tau_{n-1})$ can be any given value, whereas $u^*(x,\tau_n) = 0$. Thus the function $U^*(x,t)$ is not the approximate solution to a single adjoint system but is rather a sequence of approximations to separate adjoint systems corresponding to each interval $[\tau_{n-1}, \tau_n]$, each with the final condition

$$U^*(x,\tau_n) = 0 \quad (n = 1,2, \ldots, K)$$

for $0 \leqslant x \leqslant l$.

Exercise 3   Verify that the step-by-step solution of

$$\frac{\partial u}{\partial t} - \frac{\partial^2 u}{\partial x^2} = 0 \quad (0 < x < l;\, t_0 < t \leqslant t_1)$$

as given above, using bilinear basis functions, leads to the system of difference equations

$$\Delta_t I_x U_i^n - \tfrac{1}{3} r \delta_x^2 \{ U_i^{n+1} + 2U_i^n \} = 0, \tag{6.14}$$

where Δ_t is the forward difference operator in t; I_x, δ_x^2 are difference operators in x as defined earlier and where $r = \Delta t / (\Delta x)^2$. The region is partitioned such that $\tau_{n+1} - \tau_n = \Delta t$ $(n = 0,1, \ldots, K)$ and $x_{i+1} - x_i = \Delta x$ $(i = 0,1, \ldots, N)$. At this point, those readers who are acquainted with finite difference methods might find it useful to compare the above difference equation with a standard finite difference replacement for the simple diffusion equation, such as the Crank–Nicolson scheme. A similar comparison could be made between equation (6.12) and the standard finite difference replacements for the wave equation.

Alternative 'variational' treatments of the numerical solution of the simple diffusion equation can be found in Noble (1973) and Cecchi and Cella (1973).

6.3 SEMI-DISCRETE GALERKIN METHODS

Semi-discrete methods, mentioned briefly in Section 6.1, are techniques for by-passing variational formulations of evolutionary problems. They form the basis of the most widely used methods for such problems. In Chapter 3, variational formulations were given for the Kantorovich or semi-discrete methods for elliptic problems but this approach is certainly not essential. The starting point is taken as the weak form of the differential equation — as it was with Galerkin methods for elliptic problems. As in Chapter 3, there is no attempt to prove the equivalence of classical solutions and Galerkin solutions of the differential equation; it will be assumed that there exists a unique solution which is both classical and Galerkin.

As an example, consider the differential equation

$$\frac{\partial u(\mathbf{x},t)}{\partial t} + Au(\mathbf{x},t) = f(\mathbf{x},t) \quad ((\mathbf{x},t) \in R \times (t_0, t_1]), \tag{6.15}$$

where A is a second-order differential operator such as, in two dimensions,

$$A = -\frac{\partial^2}{\partial x^2} - \frac{\partial^2}{\partial y^2} \ .$$

A solution of (6.15) is required, subject to the initial condition

$$u(\mathbf{x},t_0) = u_0(\mathbf{x}) \quad (\mathbf{x} \in R) \tag{6.15a}$$

and the boundary condition

$$u(\mathbf{x},t) = 0 \quad ((\mathbf{x},t) \in \partial R \times (t_0, t_1]). \tag{6.15b}$$

The corresponding weak form of the problem is, in the notation of Section 3.4 that for $t \in (t_0, t_1]$

$$\left(\frac{\partial u}{\partial t}, v\right) + a(u,v) = (f,v) \quad \text{(for all } v(\mathbf{x}) \in \mathcal{H}), \tag{6.16}$$

subject to the initial condition

$$(u,v)_{t=t_0} = (u_0, v) \quad \text{(for all } v(\mathbf{x}) \in \mathcal{H}). \tag{6.16a}$$

In this model problem it is clear that — using Sobolev space notation — $\mathcal{H} = \mathring{\mathcal{H}}_2^{(1)}(R)$ and also that $u \in \mathcal{H} \times C^1[t_0, t_1]$. If a more general boundary condition is specified, then it may be necessary to modify the weak form by the addition of boundary integrals. A discussion of this method of modifying functionals is given in Chapters 2, 3 and 5.

The semi-discrete approximation U is then defined in terms of the weak form of the equation, that is, for $t \in (t_0, t_1]$

$$\left(\frac{\partial U}{\partial t}, V\right) + a(U,V) = (f,V) \quad \text{(for all } V(\mathbf{x}) \in K_N), \tag{6.17}$$

subject to the initial condition

$$(U,V)_{t=t_0} = (u_0, V) \quad \text{(for all } V(\mathbf{x}) \in K_N). \tag{6.17a}$$

For the model problem it follows that $K_N \subset \mathring{\mathcal{H}}_2^{(1)}(R)$. If the functions $\varphi_i \ (i = 1, \ldots, N)$ form a basis for the subspace K_N, the equivalent formulation of the semi-discrete approximation is that for $t \in (t_0, t_1]$

$$\left(\frac{\partial U}{\partial t}, \varphi_i\right) + a(U,\varphi_i) = (f,\varphi_i) \quad (i = 1, \ldots, N), \tag{6.18}$$

subject to

$$(U,\varphi_i)_{t=t_0} = (u_0,\varphi_i) \quad (i = 1, \ldots, N). \tag{6.18a}$$

Again for this model problem, it follows that V and U should be similar functions (of $\mathbf{x}$) so $U \in K_N \times C^1 [t_0, t_1]$ and the Galerkin approximation is of the form

$$U(\mathbf{x},t) = \sum_{i=1}^{N} \alpha_i(t)\varphi_i(\mathbf{x}). \tag{6.19}$$

If the boundary conditions are inhomogeneous Dirichlet conditions, then it would be possible to define a Galerkin approximation of the form

$$U(\mathbf{x},t) = W(\mathbf{x},t) + \sum_{i=1}^{N} \alpha_i(t)\varphi_i(\mathbf{x}),$$

where $\varphi_i \in K_N$, and $W(\mathbf{x},t)$ satisfies the boundary conditions. It follows that, as for elliptic problems, it is only necessary for V to be in the energy space and no such requirement is placed on the approximation U.

The Galerkin approximation is defined by a system of ordinary differential equations in terms of the functions $\alpha_i(t)$ $(i = 1, \ldots, N)$. It follows from (6.18) that, for the model problem, these equations can be written as

$$\sum_{j=1}^{N} \left\{ \frac{d\alpha_j}{dt} (\varphi_j,\varphi_i) + \alpha_j a(\varphi_j,\varphi_i) \right\} = (f,\varphi_i) \quad (i = 1, \ldots, N) \tag{6.20}$$

and the initial condition (6.18a) becomes

$$\alpha_j(t_0) = c_j \quad (j = 1, \ldots, N), \tag{6.20a}$$

where

$$\sum_{j=1}^{N} c_j(\varphi_j,\varphi_i) = (u_0,\varphi_i) \quad (i = 1, \ldots, N). \tag{6.20b}$$

The coefficients c_j $(j = 1, \ldots, N)$ given by (6.20b) satisfy

$$\left\| u_0(\mathbf{x}) - \sum_{j=1}^{N} c_j\varphi_j(\mathbf{x}) \right\|_{\mathscr{L}_2(R)}^2 = \text{minimum};$$

and so in certain problems it might be appropriate to replace (6.20b) by a different approximation of the original data, or possibly alter the form of the approximation to satisfy the initial condition exactly (Section 6.1).

As an illustration of this method, consider the simple diffusion equation

$$\frac{\partial u}{\partial t} = \frac{\partial^2 u}{\partial x^2} \quad (0 < x < l; t > 0),$$

subject to

$$u(0,t) = u(l,t) = 0 \quad (t \geqslant 0)$$

and

$$u(x,0) = u_0(x) \quad (0 < x < l).$$

The approximate solution is

$$U(x,t) = \sum_{i=1}^{N} \alpha_i(t)\varphi_i(x),$$

where the basis functions satisfy the boundary condition and hence

$$\varphi_i(0) = \varphi_i(l) = 0 \quad (i = 1, \ldots, N). \tag{6.21}$$

The system of equations (6.20) then becomes

$$\sum_{j=1}^{N} \left\{ \frac{d\alpha_j}{dt} d_{ij} + \alpha_j c_{ij} \right\} = 0 \quad (i = 1, \ldots, N), \tag{6.22}$$

where

$$c_{ij} = \int_0^l \frac{d\varphi_i}{dx} \frac{d\varphi_j}{dx} dx$$

and

$$d_{ij} = \int_0^l \varphi_i \varphi_j dx.$$

Although semi-discrete methods have been studied theoretically, primarily in relation to parabolic equations, they can equally well be applied to hyperbolic equations, particularly equations of the form

$$\frac{\partial^2 u}{\partial t^2} + \lambda \frac{\partial u}{\partial t} + Au = 0 \quad (\lambda > 0)$$

that represent damped mechanical vibrations. In such problems a semi-discrete approximation of the form

$$U(x,t) = \sum_{i=1}^{N} \alpha_i(t)\varphi_i(x),$$

leads to a system of second-order ordinary differential equations in the unknown functions $\alpha_i(t)$. There are *two* sets of initial conditions

$$\alpha_i(t_0) = c_i \quad (i = 1, \ldots, N)$$

and

$$\frac{d\alpha_i(t_0)}{dt} = d_i \quad (i = 1, \ldots, N),$$

the constants c_i and d_i being determined from the given initial conditions

$$u(\mathbf{x}, t_0) = u_0(\mathbf{x})$$

and

$$\frac{\partial u(\mathbf{x}, t_0)}{\partial t} = v_0(\mathbf{x}),$$

such that

$$\left\| u_0(\mathbf{x}) - \sum_{i=1}^{N} c_i \varphi_i(\mathbf{x}) \right\|^2_{\mathscr{L}_2(R)} = \text{minimum}$$

and

$$\left\| v_0(\mathbf{x}) - \sum_{i=1}^{N} d_i \varphi_i(\mathbf{x}) \right\|^2_{\mathscr{L}_2(R)} = \text{minimum},$$

respectively.

Non-linear problems

In Chapter 3 we give an example of a non-linear equation, $A(u) = f$, that can be solved by Galerkin's method. At the same time it is pointed out that the advantages of Galerkin's method are not fully realized if it is not possible to apply integration by parts to simplify the inner products. The situation is very similar when we apply the semi-discrete Galerkin method to solve the non-linear parabolic equation

$$\frac{\partial u}{\partial t} + A(u) = 0.$$

We consider, as an example, the same non-linear term as in the elliptic case, namely

$$A(u) = -\frac{\partial}{\partial x}\left(p(u) \frac{\partial u}{\partial x} \right) - \frac{\partial}{\partial y}\left(q(u) \frac{\partial u}{\partial y} \right).$$

In order to apply the semi-discrete Galerkin method to this equation it is first necessary to rewrite the equation in the weak form

$$\left(\frac{\partial U}{\partial t}, \varphi_i \right) + a(U, \varphi_i) = 0 \quad (i = 1, \ldots, N).$$

After the second term has been simplified by integration by parts, there remains a system of non-linear ordinary differential equations of the form

$$\sum_{j=1}^{N} \frac{d\alpha_j}{dt} d_{ij} + c_i(\alpha) = 0 \quad (i = 1, \ldots, N)$$

(cf. (6.22) and (6.37)), where $\alpha = (\alpha_1, \ldots, \alpha_N)^{\mathrm{T}}$. The problem of solving such a non-linear system is discussed later in the chapter.

Exercise 4   Describe the semi-discrete Galerkin method, using piece-wise linear basis functions, as applied to the equation of damped vibration of a string, defined by

$$\frac{\partial^2 u}{\partial t^2} + \lambda \frac{\partial u}{\partial t} - c^2 \frac{\partial^2 u}{\partial x^2} = 0 \quad (0 < x < l; \, t_0 < t \leqslant t_1; \lambda > 0),$$

subject to the conditions

$$u(0,t) = u(l, t) = 0 \quad (t_0 \leqslant t \leqslant t_1)$$

with

$$u(x, t_0) = u_0(x) \quad (0 \leqslant x \leqslant l)$$

and

$$\frac{\partial u(x, t_0)}{\partial t} = v_0(x) \quad (0 \leqslant x \leqslant l).$$

Exercise 5   Describe the semi-discrete method as applied to the equation

$$\frac{\partial^2 u}{\partial t^2} = \frac{\partial^2 u}{\partial x^2} + \frac{\partial^2 u}{\partial y^2} \quad ((x,y,t) \in R \times (t_0, t_1]),$$

subject to

$$u(x,y,t_0) = u_0(x,y)$$

and

$$\frac{\partial u(x,y,t_0)}{\partial t} = v_0(x,y) \quad \Bigg\} \quad ((x,y) \in R)$$

together with

$$\frac{\partial u(x,y,t)}{\partial n} = f(x,y,t) \quad ((x,y,t) \in \partial R \times (t_0, t_1]),$$

where ∂R is the boundary of the region $R = (x_0 < x < x_1) \times (y_0 < y < y_1)$ and n is in the direction of the outward normal.

Exercise 6　Verify that the semi-discrete Galerkin method applied to the linear differential equation

$$b_1 \frac{\partial^2 u}{\partial t^2} + c_1 \frac{\partial u}{\partial t} + Au = f$$

leads to a system of ordinary differential equations of the form

$$P\ddot{\alpha} + Q\dot{\alpha} + R\alpha = b$$

where P, Q and R are matrices, independent of time, such that $c_1 P = b_1 Q$.

6.4 CONTINUOUS METHODS IN TIME

As shown in the previous section, the semi-discrete method leads to a system of ordinary differential equations

$$A\ddot{\alpha} + B\dot{\alpha} + C\alpha = b, \tag{6.23}$$

subject to the initial condition

$$\alpha(t_0) = c_0, \tag{6.24a}$$

together with the additional condition

$$\dot{\alpha}(t_0) = d_0 \tag{6.24b}$$

if $A \neq 0$. Linear problems, in which the matrices A, B and C are constant matrices, have received most attention. In such cases it may be possible to obtain an exact solution of equation (6.23) by standard analytic methods.

If $A = 0$, the problem simplifies and the solution can be written as

$$\alpha(t) = C^{-1}b + \exp(-tB^{-1}C)(c_0 - C^{-1}b)$$

and can be computed (Wait and Mitchell, 1971) in terms of the solution of the eigenproblem

$$(\lambda B - C)u = 0.$$

Alternatively, if $B = 0$, $\alpha(t)$ can be computed in terms of the solution of the eigenproblem

$$(\lambda^2 A - C)u = 0.$$

Although it is possible to obtain the complete solution of the particular eigenproblem, the computation is very time consuming for a large system and so in such cases it is often preferable to approximate the solution of (6.23) by a few of the dominant components only. Such components invariably vary most slowly with respect to changes in time and correspond to the smallest eigenvalues. Individual eigenvalues together with the corresponding eigenvectors can be computed by inverse iteration (Wilkinson, 1965, p. 619) significantly faster than the complete eigensystem can be computed, so that this approach is a definite improvement provided that an approximation in terms of a few components is adequate. An approximation in terms of a few of the dominant transient components is particularly appropriate if (i) $A \neq 0$ and a *smoothing* of the oscillations is required or (ii) $A = 0$ and the *steady* state rather than the *initial* response is needed.

Consider for example the equation

$$\frac{\partial u}{\partial t} = \frac{\partial^2 u}{\partial x^2} + \frac{\partial^2 u}{\partial y^2} \quad (0 < x, y < \pi; t > 0), \tag{6.25}$$

subject to the boundary conditions

$$u(x,0,t) = u(x,\pi,t) = 0 \quad (0 \leqslant x \leqslant \pi), \tag{6.26a}$$

$$u(0,y,t) = 0 \quad (0 \leqslant y \leqslant \pi) \tag{6.26b}$$

and

$$u(\pi,y,t) = \sin y \quad (0 \leqslant y \leqslant \pi), \tag{6.26c}$$

together with the initial condition

$$u(x,y,0) = \sin y \, \frac{x}{\pi} \quad (0 \leqslant x, y \leqslant \pi). \tag{6.26d}$$

A finite element solution is required using the semi-discrete method with a bilinear basis.

As the boundary condition (6.26c) is inhomogeneous (and smooth) a transformation can be used. Thus as a comparison of the different methods, we seek an approximate solution of the form

$$U(x,y,t) = V(x,y,t) + W(x,y), \tag{6.27}$$

where

$$V(x,y,t) = \sum_{ij=1}^{N} V_{ij}(t)\varphi_{ij}(x,y) \tag{6.28}$$

is a piecewise bilinear approximation to the solution of

$$\frac{\partial v}{\partial t} = \frac{\partial^2 v}{\partial x^2} + \frac{\partial^2 v}{\partial y^2} - \frac{x}{\pi} \sin y \quad (0 < x, y < \pi; t > 0), \tag{6.29}$$

subject to homogeneous initial and boundary conditions and where either

$$(1) \quad W(x,y) = \frac{x}{\pi} \sin y, \tag{6.30}$$

or

(2) $W(x,y) = \sum\limits_{ij=0}^{N+1} c_{ij}\,\varphi_{ij}(x,y),$ (6.31)

where the coefficients c_{ij} are determined from the initial condition.

The relative performance of various methods of computing such an approximation are compared in Figure 27, the theoretical solution being

$$u(x,y,t) = \frac{\sinh x}{\sinh \pi}\,\sin y + \frac{2}{\pi}\sum_{k=1}^{\infty}\frac{(-1)^{k+1}}{k(1+k^2)^{1/2}}\,e^{-(1+k^2)^{1/2}}\,\sin kx\,\sin y.$$

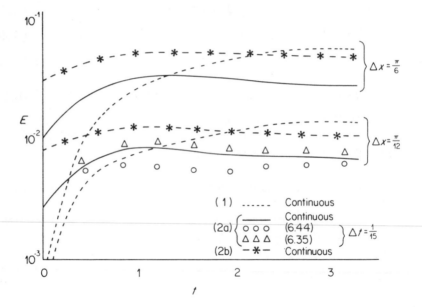

Figure 27   For method (1) $W(x, y)$ is given by (6.30); for method (2b) $W(x, y)$ is given by (6.31) and it interpolates the initial condition; for method (2a) $W(x, y)$ is given by (6.31) and it is the best $\mathscr{L}_2(R)$ approximation

6.5 DISCRETIZATION IN TIME

The closed form solutions given in the previous section cannot be applied to non-linear problems, and can rarely be applied to problems with time-dependent boundary conditions. In such circumstances, it is necessary to compute a step-by-step numerical solution. A comprehensive study of numerical methods for systems of ordinary differential equations has been provided elsewhere (Lambert, 1973) and we shall consider only methods that are suitable for computing finite element solutions. A system of equations such as (6.23) may be *stiff* (Lambert, 1973, p. 231) which means that only certain special methods will be satisfactory (Laurie, 1977; Hopkins and Wait, 1976).

Parabolic partial differential equations and the resulting first-order (in time) systems have probably received most attention and the most popular method of solution is the so-called *Crank–Nicolson–Galerkin* method. In this form of approximation, the system of differential equations

$$B\dot{\alpha} + C\alpha = b,$$ (6.32)

is replaced by the system of difference equations

$$B\left\{\frac{\alpha_{n+1} - \alpha_n}{\Delta t}\right\} + C\left\{\frac{\alpha_{n+1} + \alpha_n}{2}\right\} = b(\tau_{n+\frac{1}{2}}) \quad (n = 0,1, \ldots),$$ (6.33)

where α_n is an approximation to $\alpha(t_0 + n\Delta t)$ and $\tau_{n+\frac{1}{2}} = t_0 + (n + \frac{1}{2})\Delta t$ $(n = 0,1, \ldots)$. This form of approximate solution of a system of ordinary differential equations should be described more accurately as the trapezium method. From (6.33) it is clear that at each step of the calculation it is necessary to solve a system of linear algebraic equations to find the values of α_{n+1}. Unfortunately the situation is not so simple for non-linear equations of the form

$$B\dot{\alpha} + c(\alpha) = b$$ (6.34)

which arise from the semi-discrete Galerkin solution of equations of the form

$$\frac{\partial u}{\partial t} + A(u) = f.$$ (6.35)

A Crank–Nicolson–Galerkin approximation in the case of (6.34) would lead to a system of non-linear equations to solve for α_{n+1}, thus a *predictor–corrector* method has to be used. Douglas and Dupont (1970) suggest various schemes and discuss their relative merits, but it is only possible here to mention, briefly, a typical example of the methods they suggest for solving the equation (6.35) with

$$A(u) = -\frac{\partial}{\partial x}\left(p(u)\frac{\partial u}{\partial x}\right) - \frac{\partial}{\partial y}\left(q(u)\frac{\partial u}{\partial y}\right).$$ (6.36)

In this case $c(\alpha) = (c_1(\alpha), \ldots, c_N(\alpha))^{\mathrm{T}}$ is given by

$$c_i(\alpha) = \sum_{j=1}^{N} \alpha_j \iint_R \left\{p(U)\frac{\partial \varphi_i}{\partial x}\frac{\partial \varphi_j}{\partial x} + q(U)\frac{\partial \varphi_i}{\partial y}\frac{\partial \varphi_j}{\partial y}\right\} dx\,dy$$

$$(i = 1, \ldots, N). \quad (6.37)$$

160

Thus $c(\alpha)$ can be written as

$$c(\alpha) = D(\alpha)\alpha$$

and (6.34) becomes

$$B\dot{\alpha} + D(\alpha)\alpha = b,$$

and (6.33) is replaced by the two equations,

$$B\left\{\frac{\beta_{n+1} - \alpha_n}{\Delta t}\right\} + D(\alpha_n)\left\{\frac{\beta_{n+1} + \alpha_n}{2}\right\} = b(\tau_{n+\frac{1}{2}}) \tag{6.38}$$

and

$$B\left\{\frac{\alpha_{n+1} - \alpha_n}{\Delta t}\right\} + D\left(\frac{\beta_{n+1} + \alpha_n}{2}\right)\left\{\frac{\alpha_{n+1} + \alpha_n}{2}\right\} = b(\tau_{n+\frac{1}{2}}). \tag{6.39}$$

The predictor (6.38) gives a first approximation β_{n+1}, then the corrector (6.39) may be used iteratively to improve the approximation.

Exercise 7   Verify that the semi-discrete method, using a Crank–Nicolson–Galerkin approximation, applied to the simple diffusion equation

$$\frac{\partial u}{\partial t} - \frac{\partial^2 u}{\partial x^2} = 0 \quad (0 < x < l; t_0 < t \leqslant t_1)$$

using linear basis functions, leads to a system of difference equations

$$\Delta_t I_x U_i^n - \frac{r}{2}\delta_x^2\{U_i^{n+1} + U_i^n\} = 0, \tag{6.40}$$

where the operators Δ_t, I_x and δ_x^2, the constant r, and the partition of the region are as in Exercise 3 of Section 6.2.

One-step Galerkin methods (finite elements in time)

An alternative approach is to discretize (6.32) using a Galerkin approximation, that is, to derive an approximate solution which is of the form

$$\alpha^{(n)}(t) = \sum_{j=0}^{S} \alpha_j^{(n)}\varphi_j^{(n)}(t) \quad (n = 0,1,\ldots), \tag{6.41}$$

in each subinterval $(\tau_n, \tau_n + \Delta t)$, where the coefficients $\alpha_j^{(n)}$ $(j = 1,\ldots,S)$ are determined by

$$\langle B\dot{\alpha}^{(n)} + C\alpha^{(n)}, \varphi_j^{(n)}\rangle_n = \langle b, \varphi_j^{(n)}\rangle_n, \quad (j = 1,\ldots,S; n = 0,1,\ldots) \tag{6.42}$$

together with a continuity condition of the form

$$\alpha^{(n)}(\tau_n^+) = \alpha^{(n-1)}(\tau_n^-) \quad (n = 1,2, \ldots). \tag{6.43}$$

The ith component of $vector \langle \mathbf{u},v \rangle_n$ is

$$\int_{\tau_n}^{\tau_n + \Delta t} u_i(t)v(t)\,dt,$$

where $\mathbf{u} = (u_1(t), \ldots, u_N(t))^{\mathrm{T}}$.

Note that in (6.41), there are $S + 1$ basis functions, whereas there are only S equations in (6.42) and so the form of the difference approximation depends on the ordering of the basis functions.

We partition the subinterval $[\tau_n, \tau_n + \Delta t]$ such that

$$\tau_n = \tau_0^{(n)} < \tau_1^{(n)} < \ldots < \tau_S^{(n)} = \tau_n + \Delta t,$$

with

$$\tau_j^{(n)} - \tau_{j-1}^{(n)} = \frac{\Delta t}{S} \quad (j = 1, \ldots, S).$$

We assume that $\varphi_j^{(n)}$ $(j = 0, \ldots, S)$ form a basis for Lagrangian interpolation on $[\tau_n, \tau_n + \Delta t]$, that is,

$$\varphi_j^{(n)}(\tau_k^{(n)}) = \begin{cases} 1 & (j = k) \\ 0 & (j \neq k) \end{cases} \quad (n = 0,1, \ldots)$$

and $\varphi_j^{(n)}(t)$ are polynomials of degree S on $[\tau_n, \tau_n + \Delta t]$. From (6.43) it follows that

$$\alpha_0^{(n)} = \alpha_S^{(n-1)} = \alpha_n \approx \alpha(\tau_n) \quad (n = 1,2, \ldots)$$

and thus for $S \geqslant 2$ it is possible to eliminate $\alpha_j^{(n)}$ $(j = 1, \ldots, S-1)$ from (6.42) to give a single equation for α_{n+1} in terms of α_n $(n = 0,1, \ldots)$.

For example $(b = 0)$:

(1) $S = 1$ (Comini, Del Guidice, Lewis and Zienkiewicz, 1974)

$$\left\{ B + \frac{2}{3} \Delta t C \right\} \alpha_{n+1} = \left\{ B - \frac{\Delta t}{3} C \right\} \alpha_n \tag{6.44}$$

and

(2) $S = 2$

$$\left\{ I + \frac{3}{5} \Delta t M + \frac{3}{20} (\Delta t)^2 M^2 \right\} \alpha_{n+1} = \left\{ I - \frac{2}{5} \Delta t M + \frac{1}{20} (\Delta t)^2 M^2 \right\} \alpha_n,$$

where $M = B^{-1} C$.

It has been shown (Hulme, 1972) that several well known difference methods can be formulated as one-step Galerkin methods.

If we use Hermitian interpolation, different formulae are obtained. Another possible method of generating difference schemes, is to replace the Galerkin approximation (6.42) by a least-squares approximation.

Exercise 8   Verify that if the basis functions are placed in reverse order, i.e.

$$\varphi_j^{(n)}(\tau_{S-k}^{(n)}) = \begin{cases} 1 & (k = j) \\ 0 & (k \neq j) \end{cases} \quad (n = 0,1, \ldots),$$

and if $\mathbf{b} = 0$ and $S = 1$, we obtain the difference equation

$$\left\{ B + \frac{\Delta t}{3} C \right\} \alpha_{n+1} = \left\{ B - \frac{2}{3} \Delta t C \right\} \alpha_n.$$

Exercise 9   Different sets of difference equations can be obtained if the local approximation (6.41) is defined by

$$\langle B\dot{\alpha}^{(n)} + C\alpha^{(n)}, \psi_j^{(n)} \rangle_n = \langle \mathbf{b}, \psi_j^{(n)} \rangle_n \quad (j = 1, \ldots, S; n = 0,1, \ldots),$$

in place of (6.42), where $\{\varphi_j^{(n)}\}$ and $\{\psi_j^{(n)}\}$ are dissimilar. Verify that with $\mathbf{b} = 0$,

$$\psi_1^{(n)}(t) = 1 \quad (n = 0,1, \ldots; t \geqslant t_0)$$

and

$$\psi_2^{(n)}(t) = \frac{2}{\Delta t}(t - \tau_n) - 1 \quad (n = 0,1, \ldots; \tau_{n-1} \leqslant t \leqslant \tau_n)$$

lead to:

(1) $S = 1$

$$\left\{ B + \frac{\Delta t}{2} C \right\} \alpha_{n+1} = \left\{ B - \frac{\Delta t}{2} C \right\} \alpha_n$$

and

(2) $S = 2$

$$\left\{ I + \frac{\Delta t}{2} M + \frac{(\Delta t)^2}{12} M^2 \right\} \alpha_{n+1} = \left\{ I - \frac{\Delta t}{2} M + \frac{(\Delta t)^2}{12} M^2 \right\} \alpha_n,$$

where $M = B^{-1}C$.

Exercise 10   Verify that a least-squares approximation in place of (6.42) with $S = 1$ and $\mathbf{b} = 0$, leads to the difference equation

$$\left\{\frac{1}{\Delta t}B^2 + \frac{1}{2}[BC + CB] + \frac{\Delta t}{3}C^2\right\}\alpha_{n+1}$$

$$= \left\{\frac{1}{\Delta t}B^2 + \frac{1}{2}[BC - CB] - \frac{\Delta t}{6}C^2\right\}\alpha_n.$$

A.D.G. methods for parabolic equations (Dendy and Fairweather, 1975)

In Chapter 3, A.D.G. methods were introduced for elliptic problems. It is possible to use similar methods for parabolic problems, defined on rectangular regions. If tensor product basis functions are used, approximations defined by the *linear* equation

$$\left(\frac{\partial U}{\partial t}, V\right) + a(U,V) = (f,V) \quad (t > t_0)$$

lead (in two-dimensional problems) to an algebraic system that can be written as (see Chapter 3, p. 57)

$$A_x \otimes A_y \left\{\frac{\alpha_{n+1} - \alpha_n}{\Delta t}\right\} + (A_x \otimes B_y + B_x \otimes A_y)\left\{\frac{\alpha_{n+1} + \alpha_n}{2}\right\} = b_{n+\frac{1}{2}}$$

$$(n = 1,2, \ldots) \quad (6.45)$$

where $\otimes$ denotes tensor product. If the term $(\frac{1}{4}\Delta t)B_x \otimes B_y \alpha_{n+1}$ is added to the left-hand side, (6.45) can be replaced by

$$\left(A_x + \frac{\Delta t}{2}B_x\right) \otimes \left(A_y + \frac{\Delta t}{2}B_y\right)\alpha_{n+1} = \psi,$$

which can be solved in two stages as for elliptic A.D.G. methods (Section 3.4).

Several other methods have been proposed, in many cases attention has been restricted to linear problems; for example, the application of general multistep methods (Zlámal, 1975) and Nørsett methods (Siemeniuch and Gladwell, 1974). Dupont, Fairweather and Johnson (1974) have constructed families of three-level difference schemes for solving both linear and non-linear problems. Fairweather and Johnson (1975) have shown that it is possible to use local Richardson extrapolation based on such three-level schemes and also on certain two-level schemes proposed by Douglas and Dupont (1970). The latter authors (1975) have also considered the effect of interpolating the non-linear coefficients, that is, interpolating functions such as $p(u)$ and $q(u)$ in (6.36).

6.6 CONVERGENCE OF SEMI-DISCRETE GALERKIN APPROXIMATIONS

This section contains a brief description of one of the convergence estimates developed by Thomée and Wahlbin (1975) and Wheeler

(1973) for the model problem

$$\frac{\partial u}{\partial t} = \frac{\partial^2 u}{\partial x^2} + \frac{\partial^2 u}{\partial y^2} \quad ((x,y,t) \in R \times (t_0, t_1)),$$

subject to the initial condition

$$u(x,y,t_0) = u_0(x,y) \quad ((x,y) \in R)$$

and the boundary condition

$$u(x,y,t) = 0 \quad ((x,y,t) \in \partial R \times (t_0, t_1)).$$

The restriction to two dimensions is not significant.

The starting point for the analysis is the same as in Chapter 5; namely an assumption on the approximating properties of the subspace K_N. Thus it is assumed that Theorem 5.4 is valid and there exists $k \geqslant 1$ such that for any $u \in \mathscr{H}_2^{(k+1)}(R)$, the error introduced by interpolating u by $\tilde{u} \in K_N$, is bounded as

$$\| u - \tilde{u} \|_{r,R} \leqslant Ch^{k+1-r} \| u \|_{k+1,R} \quad (r \leqslant k). \tag{6.46}$$

If it is assumed that the variational crimes outlined in Chapter 5 are avoided, the semi-discrete Galerkin approximation satisfies

$$\left(\frac{\partial U}{\partial t}, V \right) + a(U,V) = 0 \quad (\text{for all } V \in \mathring{K}_N). \tag{6.47}$$

If $u \in \mathscr{H}_2^{\circ(k+1)}(R) \times C^1[t_0, t_1]$, it follows from (6.46) that the *projection* $W \in \mathring{K}_N \times C^1[t_0, t_1]$ which, for $t \in [t_0, t_1]$, satisfies

$$a(W,V) = a(u,V) \quad (\text{for all } V \in \mathring{K}_N) \tag{6.48}$$

also satisfies

$$\| u - W \|_{r,R} \leqslant Ch^{k+1-r} \| u \|_{k+1,R} \quad (t \in [t_0, t_1]). \tag{6.49}$$

It follows from (6.47) and (6.48) that

$$\left(\frac{\partial U}{\partial t} - \frac{\partial W}{\partial t}, V \right) + a(U-W, V) = \left(\frac{\partial u}{\partial t} - \frac{\partial W}{\partial t}, V \right)$$

for any $t \in [t_0, t_1]$; with $V = U_t - W_t$ this becomes

$$\| U_t - W_t \|_{\mathscr{L}_2(R)}^2 + \frac{1}{2} \frac{d}{dt} a(U-W, U-W) = (u_t - W_t, U_t - W_t).$$

Applying the Schwarz inequality to the right-hand side leads to

$$\| U_t - W_t \|_{\mathscr{L}_2(R)}^2 + \frac{1}{2} \frac{d}{dt} a(U-W, U-W)$$
$$\leqslant \| u_t - W_t \|_{\mathscr{L}_2(R)}^2 + \| U_t - W_t \|_{\mathscr{L}_2(R)}^2.$$

It follows from (6.49) that

$$\| u_t - W_t \|_{\mathscr{L}_2(R)} \leqslant Ch^{k+1} \| u \|_{k+1,R} ,$$

hence

$$\frac{1}{2} \frac{\mathrm{d}}{\mathrm{d}t} a(U - W, U - W) \leqslant Ch^{2k+2} \| u \|_{k+1,R}^2 .$$

and so

$$a(U - W, U - W) \leqslant Ch^{2k+2} \| u \|_{k+1,R}^2 . \tag{6.50}$$

As the bilinear form a is $\mathscr{H}_2^{\circ(1)}$ elliptic (Section 5.2), combining (6.49) and (6.50) leads to

$$\| u - U \|_{1,R} \leqslant Ch^k \| u \|_{k+1,R} .$$

This is a bound on the error in the continuous form of Galerkin approximation. If (6.47) is solved by a step-by-step procedure an additional source of errors would have to be included in the analysis. Error estimates for step-by-step solutions of Galerkin methods have been derived by several authors; additional references can be found in, for example, Dendy (1975) and de Boor (1974).

Chapter 7

Developments and Applications

7.1 INTRODUCTION

Before applying the finite element method, in its various forms, to the solution of problems, it is worth summarizing very briefly the principal characteristics of the method.

The method can be used for either steady or time-dependent problems. The finite region in space, or in space and time, is divided up into a number of non-overlapping elements. Approximating functions which may be polynomials, rationals, etc. are assumed within individual elements, and the parameters in these approximating functions adjusted so that a desired amount of continuity exists between functions in neighbouring elements. The approximating function over the complete region can then be expressed in terms of function and derivative values at nodal points in the region through basis functions which are non-zero only in a small number of elements grouped round the respective nodes. To be precise, the overall approximating function has the form

$$U(\mathbf{x}) = \sum_{i=1}^{N} \left[p_i(\mathbf{x})U_i + q_i(\mathbf{x}) \left(\frac{\partial U}{\partial x_1} \right)_i + r_i(\mathbf{x}) \left(\frac{\partial U}{\partial x_2} \right)_i + \dots \right], \quad (7.1)$$

where $\mathbf{x} = (x_1, x_2, \dots, x_m)$, the functions $p_i(\mathbf{x})$, $q_i(\mathbf{x})$, $r_i(\mathbf{x})$, etc. have local support and N is the number of nodal points in the region. The functions $p_i(\mathbf{x})$, $\partial q_i(\mathbf{x})/\partial x_1$, $\partial r_i(\mathbf{x})/\partial x_2$, etc. take the value unity at the node i. In many cases, (7.1) has the simplified form

$$U(\mathbf{x}) = \sum_{i=1}^{N} p_i(\mathbf{x})U_i, \quad (7.2)$$

although there are problems that require the more general form (7.1), particularly those where either more continuity between the elements, or greater accuracy in the gradient of the solution, is desired. The construction of the basis functions $p_i(\mathbf{x})$, $q_i(\mathbf{x})$, $r_i(\mathbf{x})$, etc. is one of the most crucial and often most difficult parts of the finite element method. This is particularly so in problems with curved boundaries and interfaces, singularities, etc. and in problems involving high-order derivatives. Chapter 4 of the present text is devoted to the construction of basis functions.

In the remainder of the chapter, a selection of problems originating mostly from physics and engineering is solved using various forms of the finite element method (e.g. Ritz, Galerkin, least squares, collocation). Many different types of basis functions and variations of the finite element method will be used in order to bring out the relative merits of the different procedures all of which are covered by the global title of finite element method. With regard to basis functions for problems where a high degree of continuity between elements is required (e.g. a least squares solution of the biharmonic equation will require C^3 continuity), non-conforming elements will be employed, and so we shall begin this chapter with a brief description of some useful non-conforming elements.

7.2 NON-CONFORMING ELEMENTS†

So far the overall approximation in a finite element method has required a certain amount of inter-element continuity. For a differential equation of order $2k$, this requirement is either C^{k-1} for Ritz and Galerkin, or C^{2k-1} for least squares. When one considers that ninthorder polynomials are required to obtain C^1 continuity between tetrahedral elements, it is obvious that for $k > 1$, the construction of elements with the required amount of continuity, i.e. conforming elements, may be a formidable task. From a computational point of view, it is thus desirable to use elements with less continuity than appears to be required, i.e. non-conforming elements.

Since engineers, of necessity, often leap in where mathematicians fear to tread, it is no surprise to learn that non-conforming elements were first introduced by engineers. A procedure called the patch test was also devised to select those non-conforming elements which lead to convergence of the finite element procedure in a given problem. In fact the patch test is a test of consistency for a non-conforming finite element method used to solve a particular problem.

The patch test (Irons and Razzaque, 1972)

A statement of the patch test is as follows: Suppose the space of non-conforming basis functions contains all polynomials of the same degree r as the highest derivative in the energy functional (in the notation of Chapter 5, $P_r \subset K_h$) and that round the perimeter of any arbitrary patch of elements boundary conditions are chosen consistent with a particular solution $u \in P_r$ within the patch. The patch test then requires the approximate solution U_h calculated by the Ritz version of

†The authors are indebted to Professor R. E. Barnhill and Mr J. H. Brown for helpful discussions during the preparation of this section

168

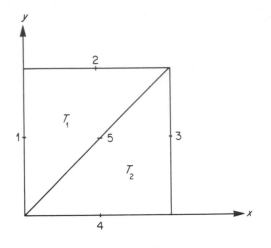

Figure 28

the finite element method ignoring discontinuities at the element interfaces to coincide with u within the patch. The patch test thus states
that if $u \in P_r$, then

$$U_h = u. \tag{7.3}$$

As an illustration of the patch test for second-order problems,
consider the solution of Laplace's equation in a unit square patch
comprising two triangular elements (Figure 28). The energy functional
contains first-order derivatives and so $r = 1$. In each triangle we choose a
linear function that matches the function values at the mid-points of
the three sides. This leads to

$$U^{[1]}(x,y) = (1 - 2x)U_1 - (1 - 2y)U_2 + (1 + 2x - 2y)U_5$$

and

$$U^{[2]}(x,y) = -(1 - 2x)U_3 + (1 - 2y)U_4 + (1 - 2x + 2y)U_5$$

in triangles T_1 and T_2 respectively. The overall interpolant is not in
general continuous across the interface of the two triangles and so this
element is non-conforming for the Ritz method.

Consider for example the test solution within the patch to be

$$u = x + y,$$

leading to the boundary values

$$u_1 = u_4 = \tfrac{1}{2}, \qquad u_2 = u_3 = \tfrac{3}{2}$$

and to the modified interpolants

$$U^{[1]}(x,y) = (-1 - x + 3y) + (1 + 2x - 2y)U_5 \tag{7.4a}$$

and

$$U^{[2]}(x,y) = (-1 + 3x - y) + (1 - 2x + 2y)U_5. \qquad (7.4b)$$

The Ritz version of the finite element method requires the energy functional

$$\iint_{T_1} (U_x^{[1]^2} + U_y^{[1]^2})dxdy + \iint_{T_2} (U_x^{[2]^2} + U_y^{[2]^2})dxdy$$

to be minimized with respect to U_5, which leads to

$$U_5 = 1.$$

Substituting this value back into (7.4a) and (7.4b) leads to

$$U^{[1]}(x,y) = U^{[2]}(x,y) = x + y,$$

and so

$$U_h = u$$

throughout the patch. In fact, this is true for any $u \in P_1$ and for any patch of elements, so this non-conforming element passes the patch test.

Exercise 1   Repeat the patch test calculation above for the elements shown in Figure 29, and show that

$$U_5 = \frac{1 - \alpha - \alpha^2 + 2\alpha^3}{1 - 2\alpha + 2\alpha^2}.$$

Hence prove that the elements pass the patch test only when $\alpha = \frac{1}{2}$.

As a further illustration of the patch test consider the fourth-order

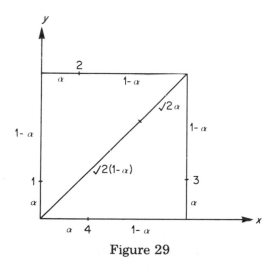

Figure 29

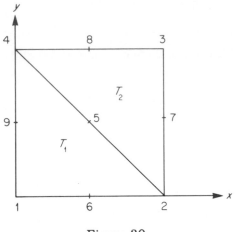

Figure 30

problem consisting of the biharmonic equation in a square region with function and normal derivative given round the boundary. The square is divided up in a standard manner into right-angled triangular elements of equal areas, and again we consider the unit square patch comprising two triangular elements (Figure 30). For the biharmonic equation the energy functional contains second-order derivatives and so $r = 2$. In each triangle we consider a quadratic which matches the function values at the vertices of the triangle and the normal derivatives at the mid-points of the sides. This is called the *Morley triangle* and leads to

$$U^{[1]}(x,y) = (1 - x - y + 2xy)U_1 + \tfrac{1}{2}(x + y + x^2 - 2xy - y^2)U_2$$

$$+ \tfrac{1}{2}(x + y - x^2 - 2xy + y^2)U_4 + y(1 - y)\left(\frac{\partial U}{\partial y}\right)_6$$

$$-\frac{1}{\sqrt{2}}(x + y - x^2 - 2xy - y^2)\left(\frac{\partial U}{\partial n}\right)_5 + x(1 - x)\left(\frac{\partial U}{\partial x}\right)_9,$$

and

$$U^{[2]}(x,y) = \tfrac{1}{2}(3x - y + y^2 - x^2 - 2xy)U_2 + (1 - x - y + 2xy)U_3$$

$$+ \tfrac{1}{2}(3y - x - y^2 + x^2 - 2xy)U_4 + x(x - 1)\left(\frac{\partial U}{\partial x}\right)_7$$

$$+ y(y - 1)\left(\frac{\partial U}{\partial y}\right)_8$$

$$+ \frac{1}{\sqrt{2}}(-2 + 3x + 3y - x^2 - 2xy - y^2)\left(\frac{\partial U}{\partial n}\right)_5,$$

in the triangles T_1 and T_2 respectively, where n is the outward normal to triangle T_1 and the inward normal to triangle T_2. Once again the overall interpolant is not in general continuous across the interface of the two triangles and so the elements are non-conforming. In a fourth-order problem the elements are still non-conforming when the interpolant is continuous across the interface but the normal derivative to the interface is not.

Consider as an example the test solution within the patch to be

$$u = x^2 + y^2,$$

leading to the boundary values

$$u_1 = 0, \qquad u_2 = u_4 = 1, \qquad u_3 = 2;$$

$$\left(\frac{\partial u}{\partial y}\right)_6 = \left(\frac{\partial u}{\partial x}\right)_9 = 0, \qquad \left(\frac{\partial u}{\partial x}\right)_7 = \left(\frac{\partial u}{\partial y}\right)_8 = 2,$$

and hence to the modified interpolants

$$U^{[1]}(x,y) = (x + y - 2xy) - \frac{1}{\sqrt{2}}(x + y)(1 - x - y)\left(\frac{\partial u}{\partial n}\right)_5 \qquad (7.5a)$$

and

$$U^{[2]}(x,y) = (2 - 3x - 3y + 2x^2 + 2xy + 2y^2)$$
$$+ \frac{1}{\sqrt{2}}(x + y - 2)(1 - x - y)\left(\frac{\partial u}{\partial n}\right)_5. \qquad (7.5b)$$

The Ritz version of the finite element method requires the energy functional

$$\iint_{T_1} (U^{[1]^2}_{xx} + 2U^{[1]^2}_{xy} + U^{[1]^2}_{yy})dxdy + \iint_{T_2} (U^{[2]^2}_{xx} + 2U^{[2]^2}_{xy} + U^{[2]^2}_{yy})dxdy$$

to be minimized with respect to the parameter $(\partial u/\partial n)_5$. This leads to

$$\left(\frac{\partial u}{\partial n}\right)_5 = \sqrt{2}.$$

Substituting this value into (7.5a) and (7.5b) leads to

$$U^{[1]}(x,y) = U^{[2]}(x,y) = x^2 + y^2,$$

and so

$$U_h = u$$

throughout the patch. In fact this is true for any $u \in P_2$ and for any patch of elements, so the Morley triangle passes the patch test.

Exercise 2 Show that the triangular element consisting of the full quadratic, interpolating the values of the function at the vertices and the mid-points of the sides does not satisfy the patch test for the fourth-order problem described above.

Although it is pleasing to verify mathematically that non-conforming elements do or do not pass the patch test, it is enough from the practical point of view to verify it on the computer. The elements are accepted as passing the patch test if the solution reproduces the exact answer, within, of course, round-off.

A necessary and sufficient condition for convergence of non-conforming elements which is equivalent to the patch test is, in the notation of Section 5.4(E),

$$a_h(p, V_h) = (f, V_h) \quad \text{(for all } V_h \in K_h) \tag{7.6}$$

for any polynomial solution $p \in P_r$, where $P_r \subset K_h$ and r is the order of the highest derivative appearing in a_h.

However any solution u satisfies

$$a(u, v) = (f, v)$$

for all admissible functions $v \in \mathcal{H}$ and *if* it also satisfies

$$a(u, V_h) = (f, V_h) \tag{7.7}$$

for non-conforming functions $V_h \in K_h$, then (7.6) can be written as

$$a_h(p, V_h) = a(p, V_h) \quad \text{(for all } V_h \in K_h) \tag{7.8}$$

(Strang and Fix, 1973, p. 178). For example if the problem is to minimize

$$I(v) = \iint_R \left[\left(\frac{\partial v}{\partial x} \right)^2 + \left(\frac{\partial v}{\partial y} \right)^2 \right] dxdy,$$

then

$$a(p, V_h) = \iint_R \left\{ \left(\frac{\partial p}{\partial x} \right) \left(\frac{\partial V_h}{\partial x} \right) + \left(\frac{\partial p}{\partial y} \right) \left(\frac{\partial V_h}{\partial y} \right) \right\} dxdy.$$

If the region R is partitioned into non-overlapping elements T_j ($j = 1, 2, \ldots, S$), it follows that

$$a_h(p, V_h) = \sum_{j=1}^{S} \iint_{T_j} \left\{ \left(\frac{\partial p}{\partial x} \right) \left(\frac{\partial V_h}{\partial x} \right) + \left(\frac{\partial p}{\partial y} \right) \left(\frac{\partial V_h}{\partial y} \right) \right\} dxdy.$$

The physical meaning of (7.8) is that the discontinuities at the inter-element boundaries can be ignored in the calculation of $a(p, V_h)$.

In order to discuss the equivalence of (7.3) and (7.8) we introduce the semi-norm

$$| u |_h = [a_h(u, u)]^{1/2}.$$

It is then possible to derive the bounds

$$| u - U_h |_h \geqslant \sup_{V_h} \left\{ \frac{| a_h(u, V_h) - a(u, V_h) |}{| V_h |_h} \right\} \qquad (7.9)$$

and (cf. (5.16))

$$| u - U_h |_h \leqslant \inf_{V_h} | u - V_h |_h + \sup_{V_h} \left\{ \frac{| a_h(u, V_h) - a(u, V_h) |}{| V_h |_h} \right\},$$

$$(7.10)$$

provided we make the assumption (7.7). Then (7.8) follows from (7.3) and (7.9). Conversely from (7.8) and (7.10), it follows that

$$| p - U_h |_h \leqslant \inf_{V_h} | p - V_h |.$$

This is zero since $P_r \subset K_h$ and so we obtain (7.3).

Exercise 3   Verify (7.7) for non-conforming piecewise linear elements that are matched at the side mid-points.

For second-order problems, a useful statement of the patch test is as follows: the patch test is passed if

$$\int_E (V_h^{[1]} - V_h^{[2]})\mathrm{d}\sigma = 0, \qquad (7.11)$$

where E is any interior straight edge of the mesh and V_h is any non-conforming function such that $V_h^{[1]}$ and $V_h^{[2]}$ are the limit values as E is approached from opposite sides (Brown, 1975).

In conclusion, we mention two non-conforming rectangular elements which pass the patch test:

(1) *The Wilson element* (Wilson et al, 1971).   On the square $0 \leqslant x,y \leqslant 1$, the six basis functions consist of the bilinear functions xy, $x(1-y)$, $y(1-x)$, $(1-x)(1-y)$ together with the additional functions $4x(1-x)$ and $4y(1-y)$. The latter functions make it possible to produce any bivariate quadratic polynomial and so permit an improved representation within each element.

(2) *The Adini element* (Adini and Clough, 1961).   This twelve-degree-of-freedom element has $u, \partial u/\partial x, \partial u/\partial y$ at the corners as unknown parameters and the complete cubic together with $x^3 y$ and xy^3 as basis functions.

Exercise 4   Verify that the Wilson and Adini rectangular elements pass the patch test for second- and fourth-order problems respectively.

7.3 BLENDING FUNCTION INTERPOLANTS

One method of deriving finite element approximations that satisfy inhomogeneous Dirichlet boundary conditions *exactly* is to include in the solution, some form of blending function interpolant of the boundary data (Gordon, 1971). The simplest case of such functions are bilinearly blended interpolants for a square.

For example, if $f \in C^{2,2}(\bar{R})$, where $\bar{R} = [0,h] \times [0,h]$, the function

$$\tilde{f}(x,y) = \left(1 - \frac{x}{h}\right) f(0,y) + \frac{x}{h} f(h,y) + \left(1 - \frac{y}{h}\right) f(x,0)$$

$$+ \frac{y}{h} f(x,h) - \tilde{\tilde{f}}(x,y), \tag{7.12}$$

where

$$\tilde{\tilde{f}}(x,y) = \left(1 - \frac{x}{h}\right)\left(1 - \frac{y}{h}\right) f(0,0) + \left(1 - \frac{y}{h}\right) \frac{x}{h} f(h,0)$$

$$+ \left(1 - \frac{x}{h}\right) \frac{y}{h} f(0,h) + \frac{x}{h} \frac{y}{h} f(h,h) \tag{7.13}$$

interpolates f exactly on the four sides of the square. In addition Gordon and Hall (1973) have shown that

$$\| D^i(f - \tilde{f}) \|_{\mathscr{L}_\infty(\bar{R})} = O(h^{4 - |i|}) \quad (0 \leqslant |i| \leqslant 1),$$

whereas the simple bilinear interpolant $\tilde{\tilde{f}}$ leads to

$$\| D^i(f - \tilde{\tilde{f}}) \|_{\mathscr{L}_\infty(\bar{R})} = O(h^{2 - |i|}) \quad (0 \leqslant |i| \leqslant 1).$$

As a numerical example of the use of blending function interpolants, we derive a finite element solution of the potential flow problem on the unit square for a source at $x = 0.437$, $y = -k$ $(k > 0)$. The theoretical solution of this problem is

$$u = \log r$$

where $r^2 = (x - 0.437)^2 + (y + k)^2$. Thus we require an approximate solution of

$$\frac{\partial^2 u}{\partial x^2} + \frac{\partial^2 u}{\partial y^2} = 0 \quad ((x,y) \in (0,1) \times (0,1))$$

subject to $u = \log r$ on the boundary.

The region is divided up into N^2 square elements and the Galerkin method is used to obtain an approximate solution that is bilinear in elements away from the boundary, but which included bilinearly blended interpolants of the boundary data in elements adjacent to the boundary. That is, the approximate solution is of the form

$$U(x,y) = W(x,y) + \sum_{j,k-1}^{N-1} U_{jk}\,\varphi_{jk}\,(x,y),\qquad (7.14)$$

where φ_{jk} $(j,k = 1, \ldots, N-1)$ are the piecewise bilinear basis functions corresponding to the points $(j/N, k/N)$, and $W(x,y)$ is a piecewise bilinearly blended function that is only non-zero in elements adjacent to the boundary and is a special case of the general form (7.12) in each such element. If the element R is of side h and has two internal nodes, and the opposite side $(y = 0)$ is part of the boundary, then in R,

$$W(x,y) = \left(1 - \frac{y}{h}\right)f(x,0),$$

since W is linear on the other three sides and zero at the internal nodes. Similarly if R is a corner element with one internal node and the sides $x = 0$, $y = 0$ are part of the boundary, then in R,

$$W(x,y) = \left(1 - \frac{y}{h}\right)f(x,0) + \left(1 - \frac{x}{h}\right)f(0,y)$$

$$- \left(1 - \frac{x}{h}\right)\left(1 - \frac{y}{h}\right)f(0,0).$$

For comparison the problem is also solved using an approximation that is bilinear in each element and which interpolates the boundary conditions only at the nodes. Numerical results are obtained for 16, 64, 144 and 256 elements for $k = 0.3, 0.2$ and 0.1. In Table 4, the maximum modulus solution on the 16 element grid is quoted in each case and compared with the theoretical solution.

Exercise 5   Repeat the calculation above for a problem involving

Table 4

No. of elements	Discre-tized	Exact	Boundary conditions Discre-tized	Exact	Discre-tized	Exact
	$k = 0.3$		$k = 0.2$		$k = 0.1$	
16	2.5747	2.5918	2.7590	2.7852	2.9756	3.0176
64	2.5875	2.5915	2.7822	2.7882	3.0200	3.0310
144	2.5896	2.5914	2.7859	2.7886	3.0281	3.0325
256	2.5904	2.5913	2.7872	2.7888	3.0306	3.0331
Theoretical solution	2.5913		2.7888		3.0339	

periodic boundary conditions where the theoretical solution is

$$u = \sin kx \ e^{-ky} \quad (k = 2.4)$$

and compare the numerical results obtained with those in Marshall and Mitchell (1973).

So far our blending function interpolants have been for rectangular elements. They can also be constructed for triangular elements and the interested reader is referred to Barnhill, Birkhoff and Gordon (1973), Barnhill and Gregory (1976a) and (1976b), and Marshall (1975).

7.4 APPLICATIONS

(A) Field problems

We now consider the solution by finite element methods of some typical boundary value field problems.

Problem 1 (Poisson's equation)

We return to a problem first considered in Chapter 3:

$$\frac{\partial^2 u}{\partial x^2} + \frac{\partial^2 u}{\partial y^2} = 2$$

in R subject to

$$u = 0$$

on ∂R where $R = (-\tfrac{1}{2}\pi, \tfrac{1}{2}\pi) \times (-\tfrac{1}{2}\pi, \tfrac{1}{2}\pi)$. The Ritz method (R.M.) minimizes the functional

$$I(v) = \iint_R \left\{ \frac{1}{2}\left(\left(\frac{\partial v}{\partial x}\right)^2 + \left(\frac{\partial v}{\partial y}\right)^2 \right) + 2v \right\} dxdy$$

whereas the Least Squares method (L.S.) minimizes the functional

$$J(w) = \iint_R \left(\frac{\partial^2 w}{\partial x^2} + \frac{\partial^2 w}{\partial y^2} - 2 \right)^2 dxdy,$$

where v and w are given by (7.1) (or (7.2)) and satisfy the boundary condition on ∂R. The Bramble—Schatz (B.S.) variation of the least squares method (Section 5.4(D)) minimizes the functional

$$J(w) = \iint_R \left(\frac{\partial^2 w}{\partial x^2} + \frac{\partial^2 w}{\partial y^2} - 2 \right)^2 dxdy + h^{-3} \int_{\partial R} w^2 \, d\sigma,$$

where the weighting factor h^{-3} is introduced to keep the dimensions of the two integrals compatible, and this time for obvious reasons w is not

Table 5

	R.M.	L.S.	B.S.
Square bilinear	0.03269		
Square cubic Hermite	0.00063	0.00063	0.06294
Triangle linear	0.03116		
Triangle cubic	0.00007	0.00283	0.02935
Triangle quintic	0.00041	0.00042	0.00321
Clough and Tocher	0.00036	0.04052	0.03752
Dupuis and Göel	0.00032	0.03395	0.03320

required to satisfy the boundary condition on ∂R. The quantity h is identified with the approximating subspace.

The region R is divided up into square elements by lines parallel to the x- and y-axes. These lines are distance h apart, where $h = \pi/N$. Triangular elements are obtained by drawing the diagonals of the squares which have slope minus one. Problem 1 is now solved for $N = 6$ by the Ritz, least squares, and Bramble—Schatz methods using a variety of basis functions. The maximum error is shown in Table 5 in each case. In general terms the least squares and Bramble—Schatz methods do rather badly in comparison with the Ritz method. This is probably due to the ill-conditioning of the linear equations which result from least squares methods.

Problem 2 (Clamped plate)

$$\frac{\partial^4 u}{\partial x^4} + 2\frac{\partial^4 u}{\partial x^2 \partial y^2} + \frac{\partial^4 u}{\partial y^4} = \frac{q}{D}$$

in R subject to

$$u = \frac{\partial u}{\partial n} = 0$$

on ∂R, where $R = (0,h) \times (0,h)$. We consider the case when the plate has a uniformly distributed load q. If w_{max} is the maximum displacement of the plate, then

$$w_{max} = \alpha \frac{q}{D} L^4$$

and the exact value of α is 0.00127. Only the Ritz method was used and for $N = 6$ the calculated values of α are given in Table 6 for a variety of basis functions.

The authors are indebted to Dr M. Vine for the numerical results quoted in Tables 5 and 6. Further details and numerical results for these and similar problems can be found in Vine (1973).

178

Table 6	
Square cubic Hermite	0.00128
Triangle cubic	0.00136
Triangle quintic	0.00127
Clough and Tocher	0.00117
Dupuis and Göel	0.00119

(B) Exact control in parabolic equations (Harley and Mitchell, 1976)

Here a finite element method is given for the problem of exact control of a linear parabolic equation. The basis functions consist of piecewise bicubic polynomials and the differential equation is satisfied at Gaussian collocation points within each element.

Consider the heat conduction equation

$$\frac{\partial u}{\partial t} = \frac{\partial^2 u}{\partial x^2} + \varphi(x,t) \quad ((x,t) \in Q = (0,1) \times (0,T)) \tag{7.15}$$

subject to the initial condition

$$u(x,0) = u_0(x) \quad (x \in [0,1]) \tag{7.16}$$

and boundary conditions for $t \in [0,T]$ consisting of

$$\frac{\partial u(0,t)}{\partial x} = f(t) \tag{7.17}$$

and either

$$\frac{\partial u(1,t)}{\partial x} = g(t) \tag{7.18a}$$

or

$$\frac{\partial u(1,t)}{\partial x} = \rho[u(1,t) - G(t)], \tag{7.18b}$$

where ρ is a constant. The function $\varphi(x,t)$ in (7.15) and $f(t)$ in (7.17) are given and it is required to find the boundary control function $g(t)$ or $G(t)$ such that the solution of the above system at some fixed time T is *exactly* $u_d(x)$, i.e.

$$u(x,T) = u_d(x) \quad (x \in [0,1]), \tag{7.19}$$

where $u_d(x)$ is a given function.

The domain Q is normalized in time to give $Q = (0,1) \times (0,1)$ and partitioned into N^2 squares of side h $(= 1/N)$ and the approximate solution $U(x,t)$ is piecewise bicubic in x and t (Section 4.2). The total number of coefficients in such an approximation is $4(N+1)^2$, but

Table 7

N^2	State function 9	16	25
$\% < h^2$	100	100	100
$\% < h^4$	100	89	80

several of the coefficients are of course known from the functions $u_0(x)$, $u_d(x)$ and $f(t)$, and if the number of unknown coefficients is $M (< 4(N+1)^2)$, then ideally in the method of collocation we require $U(x,t)$ to satisfy the differential equation (7.15) at M selected points in Q. This would provide M linear equations in the M unknown parameters. In order to collocate at Gauss quadrature points within each element, we require four collocation points per element to obtain the optimum accuracy consistent with the interpolation space used (cf. Section 3.4). This leads to $4N^2$ equations in M unknowns; an underdetermined system. This is undesirable so we choose nine Gauss points per square element leading to $9N^2$ equations in M unknowns, an overdetermined system. The latter is solved by a least squares technique and full details can be found in Harley and Mitchell (1977).

Numerical results are now presented for two problems of boundary control where the theoretical solutions are known. The control function appears in the first as part of the Neumann boundary condition (7.18a) and in the second as part of the mixed boundary condition (7.18b). The moduli of the errors in the state function and the control function are calculated at the nodes in question in the two problems and the percentages of these errors which are less than h^2 and h^4 respectively, are shown in Tables 7 and 8.

Problem 1

$$\frac{\partial u}{\partial t} - \frac{\partial^2 u}{\partial x^2} = 0 \quad ((x,t) \in Q),$$

$$u_0(x) = \sin x + \cos x,$$

$$u_d(x) = e^{-1}(\sin x + \cos x)$$

Table 8

N^2	State function 9	16	25	N^2	Control function 9	16	25
$\% < h^2$	94	98	92	$\% < h^2$	100	100	100
$\% < h^4$	60	50	40	$\% < h^4$	50	28	20

and

$$f(t) = e^{-t}.$$

What is the control function $g(t)$? (see (7.18a)). The theoretical solutions are

$$u(x,t) = e^{-t}(\sin x + \cos x) \qquad ((x,t) \in Q)$$

and

$$g(t) = e^{-t}(\cos(1) - \sin(1)) \quad (t \in [0,1]).$$

Problem 2

$$\frac{\partial u}{\partial t} - \frac{\partial^2 u}{\partial x^2} = e^{-t}\{(4x^2 - 1)\sin(x^2) - 2\cos(x^2)\}, \quad ((x,t) \in Q)$$

$$u_0(x) = \sin(x^2),$$

$$u_d(x) = e^{-1}\sin(x^2)$$

and

$$f(t) = 0.$$

What is the control function $G(t)$ when $\rho = 2$ (see (7.18b))? The theoretical solutions are

$$u(x,t) = e^{-t}\sin(x^2) \qquad ((x,t) \in Q)$$

and

$$G(t) = e^{-t}(\sin(1) - \cos(1)) \quad (t \in [0,1]).$$

(C) Numerical implementation of dual variational principles

Several examples are given in Chapter 2 of problems for which dual variational principles exist. In almost all cases numerical solutions for these problems using the finite element method have been based on the minimum principle rather than on the maximum principle. Generally speaking this is because the minimum principle is easier to implement numerically. We now solve a simple problem using a finite element method based first on a minimum principle and then on a maximum principle, and make a comparison of the results obtained.

The problem chosen is that of finding the function u which satisfies

$$\frac{\partial^2 u}{\partial x^2} + \frac{\partial^2 u}{\partial y^2} = u(x,y) \quad ((x,y) \in R) \tag{7.20}$$

and

$$u = g \qquad ((x,y) \in \partial R), \tag{7.21}$$

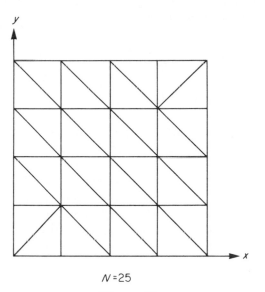

$N = 25$

Figure 31

where R is the unit square $0 \leqslant x$, $y \leqslant 1$, and g is a given function on the boundary ∂R of the square. The region is divided up into triangular elements as shown in Figure 31.

The minimum principle for this problem is

$$\underset{u}{\text{Min}} \ J,$$

where the functions u satisfy the boundary conditions and

$$J = \frac{1}{2} \iint_{R} \left\{ \left(\frac{\partial u}{\partial x} \right)^{2} + \left(\frac{\partial u}{\partial y} \right)^{2} + u^{2} \right\} \, dx \, dy. \tag{7.22}$$

If in each triangle a linear interpolant which picks up the values of u at the vertices is used, the global interpolant U takes the form

$$U(\mathbf{x}) = \sum_{i=1}^{N} U_{i} \varphi_{i}(\mathbf{x}),$$

where N is the number of nodes and $\varphi_{i}(\mathbf{x})$ ($i = 1, 2, \ldots, N$) are piecewise linear basis functions. The unknown values U_{i} at the internal nodes are obtained by minimizing (7.22) with u replaced by U.

The maximum principle (Arthurs and Reeves, 1976) is

$$\underset{U_{1}, U_{2}}{\text{Max}} \ G,$$

where $U_1 = \partial u/\partial x$, $U_2 = \partial u/\partial y$ and

$$G = -\frac{1}{2} \iint_R \left\{ \left(\frac{\partial U_1}{\partial x} + \frac{\partial U_2}{\partial y} \right)^2 + U_1^2 + U_2^2 \right\} \, dx\, dy$$

$$+ \int_{\partial R} \{ -U_2 g \, dx + U_1 g \, dy \}. \tag{7.23}$$

This form of maximum principle in terms of first derivatives is analogous to the complementary energy principle for small displacement elasticity in Section 2.6. This time linear functions are used in each triangle for U_1 and U_2. The unknown values $(U_1)_i$, $(U_2)_i$ at the nodes are obtained by maximizing (7.23) numerically.

As an example, the problem is solved using the minimum and maximum principles, where the boundary conditions are taken from the theoretical solution

$$u = e^{(x+y)/\sqrt{2}}.$$

The numerical results are as given in Table 9. Further details can be found in Freeman and Griffiths (1976).

Table 9

N	J	G
9	4.8872	4.8288
16	4.8563	4.8382
25		4.8416
Theoretical solution		4.8462

u at centre of square

N	Minimum principle	Maximum principle
9	2.0268	2.0295
25		2.0279
Theoretical solution		2.0281

(D) The critical temperature of a rod of explosive

In this example the semi-discrete form of the finite element method (Section 6.3) is used to find the critical temperature θ_{crit} of a rod of solid explosive for which one end is kept cool ($\theta = \theta_1$) and the other is kept hot ($\theta = \theta_0$). The critical temperature is such that $\theta_0 < \theta_{crit}$ leads to a steady state solution $\theta \leqslant \theta_0$ throughout the rod, whereas if $\theta_0 > \theta_{\text{crit}}$, ignition occurs at some point along the rod where $\theta > \theta_0$ and this is used as a test for ignition.

The equation governing the reaction can be written (Cook, 1958) in dimensionless units as

$$\frac{\partial \theta}{\partial t} = \frac{\partial^2 \theta}{\partial x^2} + C \exp\left(r - \frac{1}{\theta}\right) \quad (0 < x < 1,\ t > 0),$$

with C and r constants, subject to the initial condition

$$\theta(x,0) = \theta_1 \quad (0 < x < 1)$$

and the boundary conditions

$$\theta(0,t) = \theta_0$$

and

$$\theta(1,t) = \theta_1,$$

where $\theta_0 > \theta_1$.

The problem is converted into a system of non-linear ordinary differential equations by taking an approximate solution of the form

$$\theta(x,t) = \sum_{i=0}^{N+1} \alpha_i(t)\varphi_i(x),$$

where the basis functions φ_i ($i = 0,1, \ldots, N + 1$) are defined on $[0,1]$ using a uniform partition. The functions α_0 and α_{N+1} are determined by the boundary conditions and $\alpha_i(t)$ ($i = 1,2,\ldots, N$) satisfy

$$A_1 \dot{\alpha} = -A_2 \alpha + \mathbf{f}(\alpha), \tag{7.24}$$

where A_1, A_2 are positive definite matrices, a dot denotes differentiation with respect to time, and $\mathbf{f}(\alpha)$ is the non-linear vector function given by

$$f_j(\alpha) = \int_0^1 \varphi_j(x) C \exp\left(r - \left[\sum_{i=0}^{N+1} \alpha_i(t)\varphi_i(x)\right]^{-1}\right) dx \quad (j = 1,2, \ldots, N).$$

The system (7.24) is a *stiff* system of equations and it is necessary to compute the numerical solution with caution. It is solved by a variable time-step procedure devised by T. R. Hopkins, to whom the authors are indebted for the numerical results.

In the numerical test it is assumed that the solution tends to a steady state if $\theta(x,10) < \theta_0$ for $x \in (0,1)$; alternatively if $\theta(x,t) \geqslant \theta_0$ for some $x \in (0,1)$ and $t \leqslant 10$, it is assumed that ignition occurs. The critical temperature is found by the following bisection type of algorithm: if $\theta_0 = \theta_L$ leads to a steady state solution whereas $\theta_0 = \theta_H$ ($>\theta_L$) leads to ignition, then the solution for $\theta_0 = \frac{1}{2}(\theta_L + \theta_H)$ is calculated and θ_H or

184

Table 10   Linear basis functions

θ_0	0.02440	0.03050	0.02745	0.02898	0.02821	0.02859
Ignition time	∞	0.90	∞	4.74	∞	6.04
θ_0	0.02840	0.02850	0.02855	0.02857	0.02858	
Ignition time	∞	∞	∞	∞	6.5	

θ_L is replaced by this new value of θ_0 according to whether θ_0 does or does not lead to ignition.

Numerical results with various values of N are given in Tables 10 and 11 for piecewise linear, piecewise quadratic and Hermitian piecewise cubic basis functions. In all cases $\theta_1 = 0.0122$ which is equivalent to an ambient temperature of approximately $12\,^\circ$C and an ignition time of infinity indicates the solution was tending to a steady state. Piecewise quadratic basis functions are illustrated in Figure 33.

Table 11

h		¼	⅛	1/16	1/32
Linear $h = \dfrac{1}{N+1}$			0.02857	0.02843	0.02839
Quadratic $h = \dfrac{2}{N+1}$		0.02837	0.02841	0.02838	
Hermite cubic $h = \dfrac{2}{N}$		0.02822	0.02837	0.02837	

(E) Conduction—convection problems

The inadequacy of many standard finite element methods to deal with problems involving both first- and second-order derivatives of the dependent variable was first pointed out to the authors by O. C. Zienkiewicz. This is particularly so when the coefficients of the first derivatives are comparatively large. A typical example of this occurs in steady incompressible viscous fluid dynamics where the vorticity transport equation for a two-dimensional problem is

$$\frac{\partial^2 w}{\partial x^2} + \frac{\partial^2 w}{\partial y^2} - \frac{1}{\nu}\left(u\,\frac{\partial w}{\partial x} + v\,\frac{\partial w}{\partial y} \right) = 0 \tag{7.25}$$

where w is the vorticity, u and v are the velocity components, and ν is the coefficient of kinematic viscosity. The coefficients of the first-order terms in (7.25) are equivalent to the Reynolds Number and so are large in most realistic problems.

In order to illustrate the extent of the numerical problem involved in solving (7.25) we consider the one-dimensional model problem

$$\frac{d^2 w}{dx^2} - k\frac{dw}{dx} = 0 \quad (x \in [0,1]), \tag{7.26}$$

where $k = \mu/\nu$ is positive and assumed to be constant. The interval $[0,1]$ is divided into N equal parts each of length $h = 1/N$ and the nodes are located at $x = ih$ $(i = 0,1, \ldots, N)$. The boundary conditions used for the numerical examples are

$$w = \begin{cases} 1 & (x = 0) \\ 0 & (x - 1), \end{cases} \tag{7.27}$$

leading to the theoretical solution

$$w(x) = \frac{e^k - e^{kx}}{e^k - 1}. \tag{7.28}$$

The Galerkin solution W satisfies

$$(W', \varphi_i') + k(W', \varphi_i) = 0 \quad (i = 1,2, \ldots, N-1), \tag{7.29}$$

where

$$W(x) = \sum_{i=0}^{N} W_i \varphi_i(x) \tag{7.30}$$

and $'$ denotes differentiation with respect to x. If the basis functions φ_i are piecewise linear, it follows that (7.29) leads to the system of difference equations

$$(1 - \tfrac{1}{2}kh)W_{i+1} - 2W_i + (1 + \tfrac{1}{2}kh)W_{i-1} = 0 \quad (i = 1,2, \ldots, N-1) \tag{7.31}$$

This has the theoretical solution

$$W_i = A_1 + B_1 \left(\frac{1 + \tfrac{1}{2}kh}{1 - \tfrac{1}{2}kh}\right)^i \quad (i = 0,1, \ldots, N)$$

and so oscillations occur if

$$h > \frac{2}{k}.$$

Alternatively, (7.29) is replaced by

$$(W', \psi_i') + k(W', \psi_i) = 0 \quad (i = 1,2, \ldots, N-1),$$

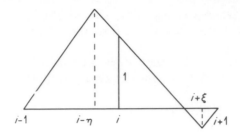

Figure 32

where W is again given by (7.30), with φ_i the conventional piecewise linear basis functions, but this time test functions ψ_i $(i = 1, 2, \ldots, N-1)$ are introduced which are asymmetric piecewise linear basis functions of the form shown in Figure 32. These were constructed by D. F. Griffiths and are given by

$$\psi_i(x) = \begin{cases} \dfrac{1 + \alpha\eta}{1 - \eta}\left(1 + \dfrac{x}{h} - i\right) & (x \in [(i-1)h, (i-\eta)h]) \\[3mm] 1 + \alpha\left(i - \dfrac{x}{h}\right) & (x \in [(i-\eta)h, (i+\xi)h]) \\[3mm] \dfrac{1 - \alpha\xi}{\xi - 1}\left(\dfrac{x}{h} - 1 - i\right) & (x \in [(i+\xi)h, (i+1)h]), \end{cases} \qquad (7.32)$$

where $0 \leqslant \xi, \eta < 1$ and α is the negative slope of the middle line. The asymmetric functions ψ_i reduce to the standard linears φ_i when $\alpha = \xi = \eta = 0$. This time the coefficients in the approximation (7.30) satisfy the difference equations

$$[1 - \tfrac{1}{2}hk(1 + \xi(1-\alpha))]\,W_{i+1} - [2 + \tfrac{1}{2}hk(\eta(\alpha+1) + \xi(\alpha-1))]\,W_i$$
$$+ [1 + \tfrac{1}{2}hk(1 + \eta(1+\alpha))]\,W_{i-1} = 0 \quad (i = 1, 2, \ldots, N-1),$$
$$(7.33)$$

leading to the theoretical solution

$$W_i = A_2 + B_2\left[\frac{1 + \tfrac{1}{2}hk + \tfrac{1}{2}hk\eta(1+\alpha)}{1 - \tfrac{1}{2}hk - \tfrac{1}{2}hk\xi(1-\alpha)}\right]^i \quad (i = 0, 1, 2, \ldots, N).$$
$$(7.34)$$

The difference equations (7.33) have first-order accuracy if $\xi \neq \eta$ and

$$\alpha = \frac{\xi + \eta}{\xi - \eta}.$$

If in addition we introduce

$$A = \frac{2\xi\eta}{\xi-\eta},$$

the theoretical solution (7.34) becomes

$$W_i = A_2 + B_2 \left[\frac{1 + \frac{1}{2}hk + \frac{1}{2}Ahk}{1 - \frac{1}{2}hk + \frac{1}{2}Ahk}\right]^i.$$

Hence there are no oscillations in the finite element solution if

(i) $A \geqslant 1$

or

(ii) $-\infty < A < 1$ and $h < \dfrac{2}{(1-A)k}$.

It should be noted that the three standard finite difference replacements of (7.26) are recovered as follows

(i) $A = 1$ (backward difference),
(ii) $A = 0$ (central difference),
(iii) $A = -1$ (forward difference).

The case $A = 0$ reproduces the Galerkin equations (7.31), which alone have second-order accuracy.

Finally, we obtain the Galerkin solution using the piecewise quad-ratic basis functions illustrated in Figure 33. The calculation leads to the difference equations

$$(1 - \tfrac{1}{2}hk)W_{i+1} - 4(2 - \tfrac{1}{2}hk)W_{i+\frac{1}{2}} + 14W_i - 4(2 + \tfrac{1}{2}hk)W_{i-\frac{1}{2}}$$
$$+ (1 + \tfrac{1}{2}hk)W_{i-1} = 0 \quad (i = 1, 2, \ldots, N-1)$$

at the integer nodes, and

$$(4 - hk)W_i - 8W_{i-\frac{1}{2}} + (4 + hk)W_{i-1} = 0 \quad (i = 1, 2, \ldots, N)$$

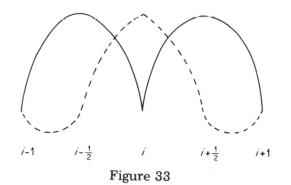

$i-1$ $i-\frac{1}{2}$ i $i+\frac{1}{2}$ $i+1$

Figure 33

Table 12   ($k = 60$)

		$h = 1/20$		
x	Theoretical	Central	Backward	Quadratic
0.90	0.9975	0.9600	0.9375	0.9941
0.95	0.9502	1.2000	0.7500	0.9231
1.00	0	0	0	0
		$h = 1/40$		
x	Theoretical	Central	Backward	Quadratic
0.90	0.9975	0.9996	0.9744	0.9974
0.925	0.9889	0.9971	0.9360	0.9885
0.95	0.9502	0.9796	0.8400	0.9490
0.975	0.7769	0.8571	0.6000	0.7742
1.00	0	0	0	0

at the half-integer nodes. These equations uncouple to give

$$(1 - \tfrac{1}{2}hk + \tfrac{1}{12}h^2k^2)W_{i+1} - (2 + \tfrac{1}{6}h^2k^2)W_i$$
$$+ (1 + \tfrac{1}{2}hk + \tfrac{1}{12}h^2k^2)W_{i-1} = 0 \quad (i = 1, 2, \ldots, N-1)$$

with the theoretical solution

$$W_i = A_3 + B_3 \left(\frac{1 + \tfrac{1}{2}hk + \tfrac{1}{12}h^2k^2}{1 - \tfrac{1}{2}hk + \tfrac{1}{12}h^2k^2} \right)^i .$$

This solution is oscillation free for all values of h and k.

Numerical results are given in Table 12, where a comparison is made of finite element solutions using linear and quadratic basis functions. Two cases of linears are quoted, viz. $A = 0$ (central differences) and $A = 1$ (backward differences). Oscillations are present with central differences for $h = \tfrac{1}{20}$. There is no question that improved accuracy could have been obtained in this particular problem for the same amount of work by taking smaller elements in the neighbourhood of the right-hand boundary. However in a problem where the form of the answer is not known, graded grids are rarely used in the first instance. Further details of this work can be found in Christie (1975).

(F) Singular isoparametric elements

In Section 4.6, it is shown that the nodes of isoparametric elements have to be placed with care to ensure that the Jacobian does not vanish within the element. There are however, circumstances under which a

Jacobian with a zero at a particular point can be an advantage. In this section we outline one application of such isoparametric elements.

If the function $u(x,y)$ satisfies

$$\frac{\partial^2 u}{\partial x^2} + \frac{\partial^2 u}{\partial y^2} = 0$$

in a region R, then *in the neighbourhood of a corner* in the boundary ∂R, it follows that u can be written as

$$u = \sum_{j=1}^{\infty} \gamma_j r^{j/\alpha} \sin\left(\frac{j\theta}{\alpha}\right) \tag{7.35}$$

for some constants γ_j $(j = 1, \ldots)$, where $\alpha\pi$ is the angle subtended by R at the corner and (r,θ) are the polar coordinates with the corner as origin. Thus it follows that near a re-entrant corner $(\alpha > 1)$, the derivatives of the leading term in (7.35) are unbounded as r tends to zero. The two most common situations are $\alpha = 2$ — a region with a slit or crack — and $\alpha = \frac{3}{2}$ — a region with a right-angled 'elbow'. The leading terms in the expansion then become proportional to $r^{1/2}$ and $r^{2/3}$ respectively. One of the main reasons for the failure of many standard numerical methods for such problems, is the inability of polynomials (in r) to represent such functions sufficiently accurately.

We now give two examples of isoparametric elements that overcome this difficulty, provided that the corner of the region is taken as a vertex of an element in which the other nodes are placed in a special way.

(1) *Quadratic elements* can be used to represent the $r^{1/2}$ behaviour. If $t_4 = \frac{1}{2}(t_1 + t_2)$, $t_5 = \frac{1}{4}(3t_3 + t_2)$ and $t_6 = \frac{1}{4}(3t_3 + t_1)$ (see Figure 17, p. 94), then the isoparametric transformation becomes

$$t - t_3 = ((t_1 - t_3)p + (t_2 - t_3)q)(p + q) \quad (t = x,y)$$

and linear functions of p and q have the necessary $r^{1/2}$ form, where r is the distance from the vertex P_3. For example along $P_1 P_3$ $(q = 0)$, it follows that

$$r^2 = (x - x_3)^2 + (y - y_3)^2$$
$$= ((x_1 - x_3)^2 + (y_1 - y_3)^2)p^4$$

and so $p \approx r^{1/2}$. The Jacobian of this transformation can be written as

$$2C_{123}(p + q)^2$$

and hence only vanishes at P_3 $(p = q = 0)$.

(2) *Cubic elements* can be used to represent $r^{2/3}$ behaviour in the neighbourhood of the node P_3. When the nodes are positioned correctly,

the transformation becomes

$$t - t_3 = ((t_1 - t_3)p + (t_2 - t_3)q)(p + q)^2 \quad (t = x,y).$$

Thus, by analogy with the quadratic case, linear functions of p and q behave as $r^{1/3}$ and hence quadratic functions have the necessary $r^{2/3}$ behaviour. In this cubic case, the Jacobian of the transformation is

$$3C_{123}(p + q)^4,$$

which only vanishes when $p = q = 0$.

Additional details of singular isoparametric elements, an illustration of their effectiveness in practical computation and generalizations of this approach to different singularities and to higher dimensions, can be found in Wait (1976).

References

Adini, A., and R. W. Clough (1961). *Analysis of Plate Bending by the Finite Element Method*, Nat. Sci. Found. Rept. G7337, Univ. of California, Berkeley.

Agmon, S. (1965). *Lectures on Elliptic Boundary Value Problems*, Van Nostrand, Princeton.

Ahlberg, J. H., and T. Ito (1975). *Math. Comp.*, **29**, 761.

Arthurs, A. M. (1970). *Complementary Variational Principles*, Clarendon Press, Oxford.

Arthurs, A. M., and R. I. Reeves (1976). *J. Inst. Math. Applics.* (to appear).

Aubin, J. P. (1972). *Approximation of Elliptic Boundary Value Problems*, Wiley, New York.

Aziz, A. K. (Ed.) (1972). *The Mathematical Foundations of the Finite Element Method with Applications to Partial Differential Equations*, Academic Press, New York.

Babuška, I. (1969). *Tech. Note BN-624*, University of Maryland.

Babuška, I. (1971). *SIAM J. Numer. Anal.*, **8**, 304.

Babuška, I. (1973). *Numer. Math.*, **20**, 179.

Babuška, I., and A. K. Aziz (1976). *SIAM J. Numer. Anal.*, **13**, 214.

Babuška, I., and M. Zlámal (1973). *SIAM J. Numer. Anal.*, **10**, 863.

Baker, G. A. (1973). *Math. Comp.*, **27**, 229.

Barnhill, R. E., G. Birkhoff and W. J. Gordon (1973). *J. Approx. Theory*, 8, 114.

Barnhill, R. E., J. A. Gregory and J. R. Whiteman (1972). 749—755 in Aziz (1972).

Barnhill, R. E., and J. A. Gregory (1976a). *Math. Comp.* (to appear).

Barnhill, R. E., and J. A. Gregory (1976b). *J. Approx. Theory* (to appear).

Berezin, I. S., and N. P. Zhidkov (1965). *Computing Methods* II, Pergamon Press, Oxford.

Berger, A. E. (1972). 757—796 in Aziz (1972).

Berger, A. E. (1973). *Numer. Math.*, **21**, 345

Berger, A. E., R. Scott and G. Strang (1972). 295—313 in *Symposia Mathematica* X, Academic Press, London.

Bers, L., F. John and M. Schechter (1964). *Partial Differential Equations*, Interscience, New York.

Birkhoff, G., M. H. Schultz and R. S. Varga (1968). *Numer. Math.*, **11**, 232.

Birkhoff, G. (1971). *Proc. Nat. Acad. Sci.*, **68**, 1162.

Birkhoff, G., and L. Mansfield (1974). *J. Math. Anal. Applics.*, **47**, 531.

Bond, T. J., R. D. Swanell, R. D. Henshell and G. B. Warburton (1973). *J. Strain Anal.*, **8**, 182.

de Boor, C. (Ed.) (1974). *Mathematical Aspects of Finite Elements in Partial Differential Equations*, Academic Press, New York.

de Boor, C., and B. Swartz (1973). *SIAM J. Numer. Anal.*, **10**, 582.

Bramble, J. H., T. Dupont and V. Thomée (1972). *Math. Comp.*, **26**, 869.

Bramble, J. H., and S. R. Hilbert (1970). *SIAM J. Numer. Anal.*, **7**, 112.

Bramble, J. H., and J. A. Nitsche (1973). *SIAM J. Numer. Anal.*, **10**, 81.

Bramble, J. H., and A. H. Schatz (1970). *Comm. P. Appld. Math.*, **23**, 635.
Bramble, J. H., and A. H. Schatz (1971). *Math. Comp.*, **25**, 1.
Bramble, J. H., and V. Thomée (1974). *R.A.I.R.O.*, **8** (R-2), 5.
Bramble, J. H., and M. Zlámal (1970). *Math. Comp.*, **24**, 809.
Brown, J. H. (1975). *Non Conforming Finite Elements and their Applications*, M.Sc. Thesis, Univ. of Dundee.
Cecchi, M. M., and A. Cella (1973). 767–768 in *Proc. 4th Canadian Congress on Appld. Mech.*
Chernuka, M. W., G. R. Cowper, G. M. Lindberg and M. D. Olson (1972). *Int. J. Num. Meth. Eng.*, **4**, 49.
Christie. I. (1975). *Conduction—Convection Problems*, M.Sc. Thesis, Univ. of Dundee.
Ciarlet, P. G. (1973a). *Springer-Verlag Lecture Notes*, **363**, 21, Berlin.
Ciarlet, P. G. (1973b). 113–129 in Whiteman (1973).
Ciarlet, P. G., and P. A. Raviart (1972a). *Arch. Rat. Mech. Anal.*, **46**, 177.
Ciarlet, P. G., and P. A. Raviart (1972b). *Comp. Meth. App. Mech. Eng.*, **1**, 217.
Ciarlet, P. G., and P. A. Raviart (1972c). 409–474 in Aziz (1972).
Clegg, J. C. (1967). *Calculus of Variations*, Oliver and Boyd, Edinburgh.
Clough, R. W., and J. L. Tocher (1965). *Proc. 1st. Conf. Matrix Methods in Structural Mechanics*, Wright-Patterson A.F.B., Ohio.
Comini, G., S. del Guidici, R. W. Lewis and O. C. Zienkiewicz (1974). *Int. J. Num. Meth. Eng.*, **8**, 613.
Cook, G. B. (1958). *Proc. Roy. Soc. London*, A **246**, 154.
Courant, R., and D. Hilbert (1953). *Methods of Mathematical Physics*, Vol. I, Interscience, New York.
Crouzeix, M., and P. A. Raviart (1973). *R.A.I.R.O.*, **7**, (R-3), 33.
Davis, P. J., and R. Rabinowitz (1967). *Numerical Integration*, Blaisdell, Waltham, Mass.
Dem'janovic, J. K. (1964). *Sov. Math. Dokl.*, **5**, 1452.
Dendy, J. E. (1975). *SIAM J. Numer. Anal.*, **12**, 541.
Dendy, J. E., and G. Fairweather (1975). *SIAM J. Numer. Anal.*, **12**, 144.
Douglas, J., and T. Dupont (1970). *SIAM J. Numer. Anal.*, **7**, 575.
Douglas, J., and T. Dupont (1971). 133–244 in Hubbard (1971).
Douglas, J., and T. Dupont (1973). *Math. Comp.*, **27**, 17.
Douglas, J., and T. Dupont (1975). *Math. Comp.*, **29**, 360.
Douglas, J., T. Dupont and M. F. Wheeler (1974). *R.A.I.R.O.*, **8**, (R-2), 61.
Dupont, T., G. Fairweather and J. P. Johnson (1974). *SIAM J. Numer. Anal.*, **11**, 392.
Dupuis, G., and J. J. Göel (1970). *Int. J. Num. Meth. Eng.*, **2**, 563.
Elsgolc, L. E. (1961). *Calculus of Variations*, Pergamon, London.
Ergatoudis, I., B. M. Irons and O. C. Zienkiewicz (1968). *Int. J. Solids Structures*, **4**, 31.
Fairweather, G. (1972). *Galerkin Methods for Differential Equations*, CSIR Special Report WISK 96, Pretoria, South Africa.
Fairweather, G., and J. P. Johnson (1975). *Numer. Math.*, **23**, 269.
Finlayson, B. A., and L. E. Scriven (1967). *Int. J. Heat Mass. Trans.*, **10**, 799.
Fix, G. J. (1972). 525–556 in Aziz (1972).
Freeman, L., and D. F. Griffiths (1976). *Complementary Variational Principles and the Finite Element Method* (to appear).
Gordon, W. J. (1971). *SIAM J. Numer. Anal.*, **8**, 158.
Gordon, W. J., and C. A. Hall (1973). *Numer. Math.*, **21**, 109.
Gordon, W. J., and J. A. Wixom (1974). *SIAM J. Numer. Anal.*, **11**, 909.
Gram, J. G. (Ed.) (1973). *Numerical Solution of Partial Differential Equations*, Reidel Publishing Co., Boston, U.S.A.
Harley, P. J., and A. R. Mitchell (1976). *J. Inst. Math. Applics.*, **18**, 9.

Harley, P. J., and A. R. Mitchell (1977). *Int. J. Num. Meth. Eng.*, **11**, 345.

Herbold, R. J., and R. S. Varga (1972). *Aeq. Math.*, **7**, 36.

Hildebrand, F. B. (1965). *Methods of Applied Mathematics*, Prentice Hall, New York.

Hopkins, T. R., and R. Wait (1976). *Comp. Meth. App. Mech. Eng.*, **9**, 181.

Hubbard, B. (Ed.) (1971) *Numerical Solution of Partial Differential Equations.* II, *SYNSPADE 1970*, Academic Press, New York.

Hulme, B. L. (1972). *Math. Comp.*, **26**, 415.

Irons, B. M. (1966). *Conf. on Use of Digital Computers in Structural Engineering*, Newcastle.

Irons, B. M. (1969). *Int. J. Num. Meth. Eng.*, **1**, 29.

Irons, B. M., and A. Razzaque (1972). 557—587 in Aziz (1972).

Jordan, W. B. (1970). *A.E.C. Research and Development Report* KAPL-M-7112.

Kantorovich, L. V. (1933). *Bull. Acad. Sci. USSR*, **5**, 647.

Lancaster, P. (Ed.) (1973). *Proc. Conf. on Theory and Applications of Finite Element Methods*, Univ. of Calgary.

Lambert, J. D. (1973). *Computational Methods in Ordinary Differential Equations*, Wiley, London.

Lascaux, P., and P. Lesaint (1975). *R.A.I.R.O.*, **9** (R-1), 9.

Laurie, D. P. (1977). *J. Inst. Math. Applics.*, **19**, 119.

Lions, J. L., and E. Magenes (1972). *Non-Homogeneous Boundary Value Problems and Applications* I, Springer-Verlag, Berlin.

Lucas, T. R., and G. W. Reddien (1972). *SIAM J. Numer. Anal.*, **9**, 341.

Marshall, J. A. (1975). *Some Applications of Blending Function Techniques to Finite Element Methods*, Ph.D. Thesis, Univ. of Dundee.

Marshall, J. A., and A. R. Mitchell (1973). *J. Inst. Math. Applics.*, **12**, 355.

McLeod, R. J. (1977). *J. Approx. Th.*, **19**, 25.

McLeod, R. J., and A. R. Mitchell (1972). *J. Inst. Math. Applics.*, **10**, 382.

McLeod, R. J., and A. R. Mitchell (1975). *J. Inst. Math. Applics.*, **16**, 239.

Mikhlin, S. G. (1964). *Variational Methods in Mathematical Physics*, Macmillan, London.

Mikhlin, S. G., and K. L. Smolitsky (1967). *Approximate Methods for the Solution of Differential and Integral Equations*, Elsevier.

Miller, J. J. H. (Ed.) (1973). *Topics in Numerical Analysis* I, Academic Press, New York.

Miller, J. J. H. (Ed.) (1975). *Topics in Numerical Analysis* II, Academic Press, New York.

Mitchell, A. R. (1969). *Computational Methods in Partial Differential Equations*, Wiley, London.

Morse, P. M., and H. Feshbach (1953). *Methods of Theoretical Physics*, McGraw-Hill, New York.

Nečas, J. (1967). *Les Methodes Directes en Théorie des Equations Elliptiques*, Academia, Prague.

Nitsche, J. A. (1971). *Abhandt. d. Hamb. Math. Sem.*, **36**, 9.

Nitsche, J. A. (1972). 603—627 in Aziz (1972).

Nitsche, J. A., and A. H. Schatz (1974). *Math. Comp.*, **28**, 937.

Noble, B. (1973). 143—152 in Whiteman (1973).

Noble, B., and M. J. Sewell (1972). *J. Inst. Math. Applics.*, **9**, 123.

Oden, J. T. (1972). *Finite Elements of Nonlinear Continua,* McGraw-Hill, New Yor

Oden, J. T., O. C. Zienkiewicz, R. H. Gallagher and C. Taylor (Eds.) (1974). *Finite Element Methods in Flow Problems*, Wiley, New York.

Oganesyan, L. A. (1966). *USSR Comp. Math. and Math. Phys.*, **6**, 116.

Oganesyan, L. A., and L. A. Rukhovets (1969). *USSR Comp. Math. and Math. Phys.*, **9**, 153.

Pian, T. H. H. (1970). *Numerical Solution of Field Problems in Continuum Physics*, SIAM—AMS Proceedings Volume 2.

Powell, M. J. D. (1973). *Conference on Numerical Software*, Loughborough.

Prenter, P. M. (1975). *Splines and Variational Methods*, Wiley, New York.

Rosen, P. (1953). *J. Chem. Phys.*, 21, 1220.

Rosen, P. (1954). *J. App. Phys.*, 25, 336.

Schechter, R. S. (1967). *The Variational Method in Engineering*, McGraw-Hill, New York.

Schoenberg, I. J. (1969). *Approximations with Special Emphasis on Spline Functions*, Academic Press, New York.

Scott, R. (1975). *SIAM J. Numer. Anal.*, 12, 404.

Serbin, S. M. (1975). *Math. Comp.*, 29, 777.

Sewell, M. J. (1969). *Phil. Trans. Roy. Soc. (London)*, A 265, 319.

Siemenuich, J. L., and I. Gladwell (1974). *Numer. Anal. Report* 5, Manchester University.

Simmons, G. F. (1963). *Introduction to Topology and Modern Analysis*, McGraw-Hill, New York.

Strang, G. (1972). 689—710 in Aziz (1972).

Strang, G., and A. R. Berger (1971). 199—205 in *Proc. American Math. Soc. Summer Inst. in Partial Diff. Equns.*

Strang, G., and G. Fix (1973). *An Analysis of the Finite Element Method*, Prentice Hall, New Jersey.

Synge, J. L. (1957). *The Hypercircle in Mathematical Physics*, C.U.P., London.

Tabarrok, B. (1973). *Proc. Conf. on Theory and Applications of Finite Element Methods*, Univ. of Calgary (Ed. P. Lancaster (1973)).

Thomée, V. (1973). *J. Inst. Math. Applics.*, 11, 33.

Thomée, V., and L. Wahlbin (1975). *SIAM J. Numer. Anal.*, 12, 378.

Vainberg, M. M. (1964). *Variational Methods for the Study of Nonlinear Operator Equations*, Holden-Day, San Francisco.

Varga, R. S. (1971). *Functional Analysis and Approximation Theory in Numerical Analysis*, SIAM Publications, Philadelphia.

Vine, M. (1973). *Applications of the Finite Element Method to Partial Differential Equations*, Ph.D. Thesis, Univ. of Dundee.

Vulikh, B. Z. (1963). *Introduction to Functional Analysis*, Pergamon Press, London.

Wachspress, E. L. (1971). Conf. on Appl. Num. Anal., Dundee, *Springer-Verlag Lecture Notes in Math.*, 228, 223.

Wachspress, E. L. (1973). *J. Inst. Math. Applics.*, 11, 83.

Wachspress, E. L. (1974). Conf. Num. Soln. Diff. Equns., Dundee, *Springer-Verlag Lecture Notes in Math.*, 363, 177.

Wachspress, E. L. (1975). *A Rational Finite Element Basis*, Academic Press, New York.

Wait, R. (1976). *Singular Isoparametric Finite Elements* (to appear).

Wait, R., and A. R. Mitchell (1971). *J. Inst. Math. Applics.*, 4, 241.

Washizu, K. (1968). *Variational Methods in Elasticity and Plasticity*, Pergamon Press, London.

Watson, G. A. (Ed.) (1974). Conf. Num. Soln. Diff. Equns., Dundee, *Springer-Verlag Lecture Notes in Maths.*, 363.

Watson, G. A. (Ed.) (1976). Conf. Num. Anal., Dundee, *Springer-Verlag Lecture Notes in Maths.* (to appear).

Wheeler, M. F. (1973). *SIAM J. Numer. Anal.*, 10, 723.

Whiteman, J. R. (Ed.) (1973). *The Mathematics of Finite Elements and Applications*, Academic Press, New York.

Whiteman, J. R. (1975). *A Bibliography for Finite Elements*, Academic Press, London.

Whiteman, J. R. (Ed.) (1976). *The Mathematics of Finite Elements and Applications*, Academic Press, New York (to appear).

Wilkinson, J. H. (1965). *The Algebraic Eigenvalue Problem*, O.U.P., Oxford.

Wilson, E. L., R. L. Taylor, W. P. Doherty and J. Ghaboussi (1971). *Univ. of Illinois Symposium*.

Yosida, K. (1965) *Functional Analysis*, Springer-Verlag, Berlin.

Zienkiewicz, O. C. (1967). *The Finite Element Method in Structural and Continuum Mechanics*, McGraw-Hill, New York.

Zienkiewicz, O. C. (1971). *The Finite Element Method in Engineering Science*, McGraw-Hill, New York.

Zlámal, M. (1973). *SIAM J. Numer. Anal.*, **10**, 227.

Zlámal, M. (1974). *SIAM J. Numer. Anal.*, **11**, 347.

Zlámal, M. (1975). *Math. Comp.*, **29**, 350.

Index